# Rooted

'*Rooted* is more than a memoir; Langford n[...]
the whole scale of the coming agricultura[...]

'Shot through with tenderness and admiration for the farmers she
meets, Langford's enthralling book is an unignorable call to understand
the challenges facing not only farming but the Earth itself . . .
A beautiful and rousing book' *Spectator*

'Heartbreaking and hopeful, this story of a farming revival has never been
more important. It opened my eyes and touched my soul' Esther Freud

'Absorbing, compassionate . . . Galvanising' *Guardian*

'Sarah Langford's book on farming is really a book about healing.
All of life and death is here: family, politics, nature, climate, history,
humanity. *Rooted* is a beautifully written, powerful reminder of where
we've gone wrong, what is at stake, and how we can change. I loved it'
Christie Watson, bestselling author of *The Language of Kindness*

'Langford writes so movingly of the countryside and its effect
on her heart and her family that she makes a case almost without
arguing for the importance of these landscapes and, by extension,
the economies that support them' *TLS*

'*Rooted* offers us an honest look at the farming life today. It is not an
easy way to make a living, but through Langford's personal story – and
those of who she meets – we appreciate how it offers a connection
with the land, and a firmer sense of our place in the world. Raw, earthy
and inspiring' Cal Flyn, author of *Islands of Abandonment*

'A refreshing perspective on an overwhelmingly masculine
world . . . The stories it contains coalesce into a powerful
narrative of struggle and innovation' *Financial Times*

'A beautifully written, incredibly timely book addressing not just where
our food comes from and why this matters so much, but also fundamental
questions relating to our relationship with the land, and the definition of
home' Clover Stroud, author of *My Wild and Sleepless Nights*

'*Rooted* is a brave thing: a book that prods into the ever-widening
gulf between the binaries we increasingly use to examine the world.
As conversations about what we eat and where it comes from reach
fever-pitch, Sarah Langford's clear-eyed, inquisitive and passionate plea
for farmers and farming offers a vital understanding when it has
never been so needed' Alice Vincent, author of *Rootbound*

'Moving, startling, uplifting, galvanising and unsettling, this plainly beautiful book is one of those rare few that changes how you see the world around you: the shape of fields seen from a train, the vegetables in a supermarket chiller cabinet, the earth beneath your feet and falling through your fingers. I needed to read this' Ella Risbridger, author of *The Year of Miracles*

'An eloquent and personal insight into the terrible human as well as environmental cost of cheap food and an inspiring account of the people working to heal our relationship with our habitat and ourselves. Urgent, necessary and moving' Ben Rawlence, author of *The Treeline*

'A fine book: heartfelt, honest and hopeful. Sarah has the knowledge and skill to help people better understand where their food comes from and why we should all care' Helen Rebanks

'A magical book of wonderful stories about how farmers think and the challenges they face. It demonstrates that farmers across the country are passionate about producing food and caring for the land. A triumph' Jake Fiennes, author of *Land Healer*

'Heartbreaking and heartwarming, these human stories refract the eye-watering truth of real life and the vital taproot of farming, shedding light, darkness and all the hard bright blinks of weather and season that fall between. Moving, intimate, tender and searing, this is a gem of a book with deep roots and fresh green shoots. I want to devour everything else she has written' Tamsin Calidas, author of *I Am An Island*

'A timely and optimistic book, ostensibly about why we need farming to produce food, but more deeply about *how* farming is done, or could be done. Refreshingly authentic, *Rooted* gives us a hopeful sense of a regenerative future' Juliet Blaxland, author of *The Easternmost House* and *The Easternmost Sky*

'Evocative and resonant. These are stories that need to be told' Andy Cato, Groove Armada and Wildfarmed

'Excellent [and] immensely readable . . . Where *Rooted* ploughs its own shining furrow is in its humanity, the cost to farm folk of conventional agriculture, but also the gathered, inspirational stories of farmers trying to do better and greener' *Country Life*

# Rooted

*How Regenerative Farming
Can Change the World*

SARAH LANGFORD

PENGUIN BOOKS

PENGUIN BOOKS

UK | USA | Canada | Ireland | Australia
India | New Zealand | South Africa

Penguin Books is part of the Penguin Random House group of companies
whose addresses can be found at global.penguinrandomhouse.com.

First published by Viking 2022
Published in Penguin Books 2023
004

Copyright © Sarah Langford, 2022

The moral right of the author has been asserted

Typeset by Jouve (UK), Milton Keynes
Printed and bound in Great Britain by Clays Ltd, Elcograf S.p.A.

The authorized representative in the EEA is Penguin Random House Ireland,
Morrison Chambers, 32 Nassau Street, Dublin D02 YH68

A CIP catalogue record for this book is available from the British Library

ISBN: 978–0–241–99182–4

www.greenpenguin.co.uk

For all the cranks who caused a revolution.

And in memory of my grandparents,
Peter and Dina Flindt.
For all you taught me, even if you never knew it.

# Contents

# Introduction

*I go down to the shore in the morning*
*and depending on the hour the waves*
*are rolling in or moving out,*
*and I say, oh, I am miserable,*
*what shall —*
*what should I do? And the sea says*
*in its lovely voice:*
*Excuse me, I have work to do.*

Mary Oliver

This story happened by accident. I never expected, or wanted, to live a life on the land. But then fate took me there and taught me lessons I never knew I needed to learn.

Some people tell me that having farming in my family means it must be in my blood. Maybe they are right. This book is, in part, my journey to find out. First, I had to decide what being a farmer meant. To do that I had to find out what being a farmer used to mean, what it now means and what it will mean in the future. In doing so, I met others asking themselves the same questions. This book weaves their stories around mine. Some of those farmers — like Tom, Rebecca and Stuart — are still discovering the answers. Others – like Ollie, Sam and George – found them long ago. Like many of the farmers in this book, to find my future I had to go back to my past. So that story – the story of my grandparents' farm – is the first that you will read.

As I made this journey, I came across a revolution: a way of farming that is both new and also very old, and which asks us to look at our history, our future and our values differently. It is a revolution that might just abate a climate crisis, a physical and mental health crisis, and a biodiversity crisis too. It could help to regenerate our land while also giving us some of the answers so many city dwellers seem to be looking for, not just about connection but about how to live. In doing so, it can teach us a different way of being.

I found myself in this world at the beginning of one of the greatest upheavals in farming for generations. Brexit has separated the United Kingdom from Europe, and so too from the Common Agricultural Payments that have supported food production for so long. Now these public subsidies are being withdrawn. One third of all UK farms make a loss without them.[1] Farmers will now be granted public money for looking after land that many accuse them of wrecking, and not for growing the food they define themselves by. Food is no longer seen as a public good, but as a public given. This shift in mission is both huge and difficult, because farming is more than just a job. It is an identity. Everything about it must be seen through this prism. It is not just how farmers spend their days that is changing. It is who they are.

Because the devolved nations of the United Kingdom are planning their futures slightly differently, I have focused on English farmers in this book, although farmers all over the world are facing the same battles – mental, financial and environmental – that lie within these stories. The change in government policy has made English farmers a test case: how we define farming now may alter the way others come to see it around the globe.

Some of the farmers in this book do not want a part in public life save for their stories to be better understood, so I have changed their names and identifying details to protect their anonymity. However, the people in this book are real and so are their stories, their views, their experiences, their heartbreaks and their joys. I use each farmer's exact words where I am able, and what I know of them, their lives, their expressions and opinions to narrate the rest where I am not.

It is important to be clear that this book is not an anthology of farming. It cannot be. Every farmer will do something different from their neighbour. But while their methods and produce may change, the themes that run throughout their lives and their farms are often the same. This journey has taken me around the country speaking to farmers from every walk of life who are farming in many different ways, on varying soils and on land with diverse histories. They have trusted me with their diaries, old correspondence, home movies and photographs. They have confided moments of pain and joy. More than one has cried at the relief of unburdening themselves. I have learned a lot from every one of them.

But this journey has meant more than the gathering of knowledge. It has involved an understanding, not just of the world around me – although it certainly taught me to look at things differently – but also of myself. I had placed a learned emphasis on certain things and on certain people. I ended up in the countryside seeking a break from reality. Instead, I ended up finding it.

*You have to take a view that everything*
*happens for a reason.*
*You wouldn't be doing what you're*
*doing now if that hadn't happened.*

Rebecca, farmer

# 1. The Beginning – Butter and Honey and Dust

It is the smell that makes me think of him: butter and honey and dust. It comes from the meadow field, which is separated from our cottage garden by a wooden fence. I lean against the fence and inhale. In a few weeks' time these meadow grasses will be cut. They will be bundled into great cylinders, each one rolling out of the red and rusty baler. I will shout to my two young sons to come and see its cleverness. Wilfred, nearly three years old, will run barefoot, whippet fast, and behind him my husband, Ben, will carry our ten-month-old baby. Together we will line up against the fence and watch. Each time the machine stops, quivers, and births a bale we cheer.

Now, though, on this evening in the east of England, the scent of the uncut hay hangs in the air. Behind me, inside the cottage we have just moved into, are bags and boxes waiting to be unpacked. We have only brought essentials with us. We don't plan on being here long. But I don't want to go inside. I want to be here on the edge of a meadow, wondering about this new life we are about to make, which looks and feels so very different from the one we have just left.

Wilfred climbs up the broken gate in the middle of the fence and sits on it, his legs either side of the top bar. His brother pulls himself up on the lower rung, grizzling because he is still too small to do the same. They are clean and pink from their bath. I look at them and try to remember how it feels to be that small, with bare feet and limbs and nostrils filled with the scent of butter and honey and dust.

I close my eyes to better see the man who belonged to that smell. I can feel the sensation of hugging him as clearly as if it had just happened, his tweed jacket rough against my cheek. My grandfather was an oak of a man: six foot four inches tall with huge farmer's hands and a distrust of deodorant. He smelled of body and masculinity and of the earth. His face had seen all the seasons, brown as brick, cracked with lines. He had a thick black moustache and his trouser waist was always high; his back always straight. Like others of his time and occupation he was, I think, suspicious of sentiment. After he died we discovered dozens of small notebooks into which he had occasionally written life events alongside his records of the Hampshire farm where he lived and worked. On the day my mother, his eldest child, was born, he wrote of my grandmother, 'Dina calved. Heifer calf.'

But I also remember what his face looked like each time he saw me. I was his eldest grandchild from his eldest child, and when he sat down I would climb up his giant frame to make him read me stories. The ones he read had rabbit heroes, even though he loved to lean out of his farmhouse bedroom window and shoot the real ones to death. Once, long before I was born, he nearly hit a rambler too, missing him by a few inches and sparing himself a charge of manslaughter. But the land the man was walking on was not my grandfather's to *get off*. It never would be. My grandfather was a tenant farmer. The fields he worked and the home he lived in were rented. They did not belong to him and never would. What he did, how he lived, his work and legacy, had to be enough for a lifetime's security and achievement and pride. And, back then, it was.

My grandparents' farm and the freedoms it offered me and my two younger sisters punctuated our lives. With three children aged under four and the farm less than twenty minutes' drive away, my mother sought out its refuge often. It was the backdrop to nearly every weekend and school holiday. Once, for several months, it became our home. Farming wove its way around my childhood not just because of the farm and the farming friends my mother had grown up with but because of my father's job as a land agent. As a child, all I understood of his job was that he spent his days talking to farmers and landowners and kept wellingtons and a wax jacket in the boot of his car, which made it smell always of the outdoors. But, without my really noticing it, the markers of the rural calendar became the markers of my childhood. Agricultural shows, harvest teas, point-to-points, shoots. Ploughing time, planting time, lambing time, harvest time. Bullocks and bulls; calves and lambs. Spring cuckoos; blackberry picking; winter sloes.

This world was familiar, and I therefore took its education and its access entirely for granted. It was only when I grew to know those who had not had it that I truly understood both my luck and its gifts. That is to say,

I grew up unafraid of the countryside and the people in it. I was not afraid of the animals or of the machines. I was not afraid of nature, red in tooth and claw. I was not afraid of either death or killing, by which I mean I was – mostly – not afraid of life, for they are both part of the same circle. But only now, when I am standing at the edge of a Suffolk meadow, do I understand that all these fragments of rural life have collected together to make something substantial. They have formed an imprint, a blueprint, waiting for me to need it.

Now I do need it. We are living next to a meadow by accident. I had no dreams of a life on the land. I had loved my life because it was in the city, and I love cities. I was born at the beginning of the 1980s, a decade defined by individualism, commercialism, capitalism, The City. I knew I wanted to be one of the women in the films I watched, always marching down a crowded pavement swinging a briefcase, wearing stilettos and shoulder pads. I wanted to be what that image meant: clever and powerful and important. Everything I saw and heard told me a city was the only place with people worth listening to. I had absorbed a truth that said, for your life to matter, for it to be of value, for your voice to be heard, you could not live where people's voices were soft. I moved to London as soon as I could.

I loved how arrogant and fast and extreme and anonymous it was. I would wonder at how, at any given moment, even when I was asleep, there was someone just metres away from me whom I would never know. City life had a thousand different distractions: the current of life was turned up to full wattage, day and night. Like disco lights, there were people always streaming past to catch my eye, even if none of them looked back. City air hung heavy with urgency. Sometimes I felt that it charged me up. My every sense – smell, sight, hearing, taste, touch – was stimulated, so that I learned to dull my perceptions a little in order not to be overloaded. It took a while for me to realize that it could sometimes be hard, after so much noise, to really hear.

I married, had children and put my work as a criminal and family barrister on hold to care for our new baby and toddler, and to write. In 2017 we bought our first family home, an unloved wreck in south London with missing ceilings and stairs which had long been left empty, marketed by estate agents as desperate as we were. But then, at the start of the summer, Ben lost his job and our life swivelled on its axis. We decided to move out of London as we made our new, uninhabitable home safe to live in. We would go to Suffolk, where Ben grew up, and rent a cottage near his parents. I could finish my first book and we could try to work out what to do next. It would be a respite from upheaval. A break from real life.

We packed up the place in London that we had been renting and put its contents into storage. Wedding-present china. Coffee-table books. Lawyer suits. Candlesticks. The Nespresso machine. A glass-topped dining table. The debris of modern urban life was interned in a colossal warehouse in north London. We told the storage company we would come and collect it in six months when we moved back. Except we won't. For, as though to make certain of the impossibility of doing so, the storage facility then burned down with our things inside it. The fire lasted for three days. It took 120 firefighters to put it out. Fate clearly wanted the job done properly.

In six short weeks the life I had been leading for so long has been upturned. Now I am standing in the garden of our new home wondering why I am there.

In front of me the meadow field and the two other pasture fields that lie beyond it spread out in an echo of their early pre-eighteenth-century enclosure boundary. Our cottage lies in a valley full of gentle slopes in one of England's flattest counties, cut through with a river. The land here is arable country: filled with crop fields of many hundred acres and soils that range from free-draining sandy loams to heavy clay. The three pasture fields in front of me belong to Ben's family, bought up bit by bit over the last ten years. Together with three crop fields nearby, they form a small farm that has, up until now, been managed at one remove by owners rather than farmers. But now we are here. Jobless. A little lost. A little bruised. So while we work things out we will take on the running of this farm, just for a few months while I finish writing my first book and Ben looks for other work. Afterwards, I wonder whether there is another reason we ask to do it. Looking after this land might give us a way to feel grounded when life as we know it has been uprooted.

The sky begins to slash into ribbons of pink and I

realize how late it must be. I pick the children up, one under each arm, cross and squirming because they don't want to go inside either. After I have settled them into bed I leave them with Ben, then walk back out of the cottage and into the garden alone. I feel like I need to be outside. I need to try and think and better work out why we are here and what the future holds.

The grass and clover feel cool under my feet. The air is filled with the low bass line of distant combine harvesters and the sound of birds chattering their way to bed. Butterflies and tiny brown and white moths dance over the tips of the meadow's long grass. I lean against the fence and watch a gang of swallows dart and skim and dive, arrows of blue-black, the white of their breast feathers flashing at me as they pitch and swerve to snatch insects in mid-air. They are so fast I think they must be swifts, because I have not looked carefully at either bird since I was a child and it is only later that I begin to learn how to tell the two apart. Standing, watching them, another memory returns. Sitting on the grass outside the back of my grandparents' farmhouse on a summer's evening like this. My grandmother shelling peas, grown by my grandfather, into a stoneware bowl. I sit next to her. Above us comes the *scree, scree* call of swifts. My grandmother points to the sky as they dart around the eaves of the farmhouse, in the same place bats would fly in the evening. I try to see them but they are a blur: too high and fast. Over a lifetime those swifts will have flown far enough to have made it to the moon and back five times. They have eaten, drunk, mated and slept in flight, a fact so beyond comprehension that it was not until an airman during the First World War shut off his engine 10,000 feet above the ground to glide silently above the enemy, and then found himself amidst the sleeping birds, that we believed it.[1] I watch the swallows skim the meadow fields, still then unsure if they are swifts or not, and I think of how the life I have just left behind could sometimes feel like that. Eating, drinking and mating in flight, my feet never touching the ground. Until now.

We have moved from somewhere that felt like the middle of everything to somewhere that feels like the middle of nowhere. We have been blown into this new life by a storm we had not seen coming. The storm has plucked us up like Dorothy in *The Wizard of Oz* and spat us out on the edge of a field into a life very different from the one we had before. The news is full of snap election results, of Brexit, of Trump, of Russian interference in elections, of North Korean missile testing and Syrian chemical weapons, of London terrorist attacks, of Grenfell burning. It feels like the world is being torn down its seams at the same time as our family life has been. But this new life has an echo of one I had once known. Except – like the Scarecrow, Lion and Tin Man – the people, machines and landscapes in it are both very familiar yet also completely changed from how I remember them.

On the horizon, the wide-open sky washes twilight blue. In front of me, the meadow field slopes gently up, then down and away again, rising on the other side into a line of trees. Now, in the dusk, their silhouettes look like a woodcut. From where I now stand I can see the combine harvester in the field on the other side of the valley, divided from us by river and road. I wonder who is driving the combine, and whether they own the land they are working on, and, if not, whether they even know who does. I think of how, in the city, we hold two contrasting pictures of a farmer: one from a children's picture book and one from a poster of ecological destruction. I wonder whether either of these cartoons is true any longer. I wonder if anyone knows what being a farmer means any more. I need to find out because now, unexpectedly, I have been given the chance to become one and I am starting to think I might want to take it.

As I watch the combine cross back and forth, I am struck by the simple truth that farmers have worked and shaped the land I'm looking at for thousands of years. Beyond these fields it might feel as though the planet is beginning to fall apart. Our own small world may feel like it's already fallen apart. But, as I stand and watch, I feel something reassuring in the thought that crops will be sown, crops will grow and crops will be harvested, no matter what chaos lies beyond them. I am about to become a small link in this long chain. The decisions and actions we will take on this land will bind us not just to those who will come after, but also to those who came before. It is not my land and it is not the place that I grew up, but maybe, I think, this does not matter. Maybe I can learn this land and grow roots here. I do not know then that, in fact, the method will work the other way round. The land will teach me about myself and, somehow, it will grow itself into me. Something that I thought was an end is about to turn into a beginning.

I stare out at the meadow field thinking about all that has passed and all that is to come. Then, suddenly, from the edge of my sightline, a white barn owl swoops. It is huge, its wings stretched wider than I ever knew they could be. I instinctively reach for my phone to photograph it then realize I have left it inside. The owl floats noiselessly towards and away from me, towards and away again, then returns into the copse of trees in the far left corner of the meadow. I wait, holding my breath. Then it flies back out and straight at me, close enough for me to think I might be under attack. When it is a few metres away it pulls to a halt, pulses in the air for a few wing beats, then throws itself down into the meadow. I can see only the glowing white of its back fringed by tall grasses. When it rises, its small fat quarry is clamped in its talons. It lands on a fence post only a few metres to my left. I keep my body as still as I can and turn my head to watch it. The sound of cracking bones rings out into the evening air as it finishes its kill. Then it turns and looks straight at me with its whole head, its dark eyes locked on mine. My ears rush with the noise of my heart thumping blood around my body. I feel as if I cannot breathe. Being seen by this creature feels like being allowed into a secret world. The sensation of it makes my skin prickle. Then, in one motion, the owl takes flight back towards the copse, its prey in its talons and the evening sky behind it.

I let out my breath over the meadow and with it more

than the novelty of what I have just seen. I breathe out the uncertainty of our life, my worry about the stability of our own small family and our small country. I inhale the smell of butter and honey and dust and wonder what my grandfather would think about the fact that I now find myself not in a courtroom, but on the land.

I don't yet know what the future holds for this farm in Suffolk, nor what it holds for my grandparents' farm either. My uncle, who has taken over their tenancy and now works the land, finds himself with an inheritance very different from my grandfather's. I have read that over 10,000 farms will go under in the next ten years. His farm – my grandparents' old farm – may be one of them. If it is, I wonder what parts of me will be lost with it, for that farm helped form the blueprint of who I am, and it is this blueprint that might just enable me to sink new roots into new land on the other side of the country from where my grandfather once sank his own.

My grandfather, Peter, was considered a hero who fed a starving nation. Now his son Charlie, my uncle, is considered a villain, blamed for ecological catastrophe and with a legacy no one wants. The story of their farm is also the story of farming. It tells us about farming's past and how and why it changed in just a few short decades into something so very different. But their story might tell us something else too. It might tell us the story of farming's future.

*These are complicated stories,*
*they are nuanced stories,*
*and they are being told in a binary world.*

Sarah, farmer

# 2. Peter and Charlie

The field is small by modern standards, just twelve acres. It is called Englands, which is fitting, for its size and shape mean it could belong to few other countries than its namesake. At its base lies a small copse, which, in autumn, burns every shade of red, yellow, orange and brown. The colours of its leaves are so bright that they can be seen like a torch flame from the bedroom window at the front of the farmhouse opposite.

The field is the start of a farm that rolls down into the valley, its boundary shaped like a boot. Nine hundred and twenty acres of land split into thirty fields, two woods, three cottages, a yard, barns, a red-bricked farmhouse and a road that runs down the farm's middle. Not one of its fields is square and few are flat. Some who know it say this is difficult land – 'bloody difficult land' – to farm. Its soil can turn from sticky clay in the morning to dry and impenetrable earth by the afternoon. In a wet autumn some fields will lie waterlogged for weeks. Flints work their way to the surface with such regularity that the men who once laboured here swore that the stones must be alive, growing up out of the soil like the crops they harvested. It harbours several Bronze Age burial barrows, the source of a river and the debris of farming's history – ploughshares, horseshoes as big as dinner plates – buried in its soil. It is a place so special that, when I am at school and asked to write about somewhere important to me, it is this farm that I choose.

It was September 1959 when the farm tenancy last changed hands from one family to another. It happened, as it always does, on the 29th of the month at Michaelmas: one of the quarter days that fall close to a solstice or equinox and divide the year into four. Michaelmas marked the end of the season of bounty and the beginning of a new cycle of farming, and so it became the date for new beginnings: for legal, school and university terms, when new magistrates were elected, servants hired and rents became due. And so it became the date new farm tenancies began.

On that Michaelmas Day in 1959, the new tenant, Peter, drives up the steep and narrow hill that leads from the road to the farmhouse, sweeping his car left then right. He passes cottages, a field, then reaches a long red-brick wall on his left and, on his right, a sloping field with a copse burning shades of crimson at its base. At the end of the wall he stops, turns sharply left and drives into the farmyard. Later that day he will write in his diary, 'Afternoon took car to garage. Got Land Rover. Evening to Hinton.'

Peter needs a Land Rover because this is now his farm. He had been born to farm, but born without one. As a boy he spent the holidays away from the suburbs of his childhood with an unmarried aunt, whose farm he fell in love with. By his teens he had a morning milking round in a dairy. By the time the war began to grumble, a farming apprenticeship. In the late summer of 1939, when Michaelmas was near and autumn had arrived again in her layers of red and gold, he watched the boys and men go off to war but knew he had a different calling. His war work must take place in a field without a battle in it.

This tenancy, then, is a dream now realized. Each part of this farm will become his, and he will become part of it in every way but law, for he will never own it.[1] The farmhouse and land around it belong to the squire who lives in the big house with the towering metal gates, which rules over its domain from the top of the valley.

For twenty-five years this farm has been worked by a man whose height is as sizable as his stomach. Doug Reid, six foot two inches tall and dressed in leather gaiters and an overcoat with odd buttons, lives in only a few of the rooms in the farmhouse, unmarried and child-free. He has been looked after by a local housekeeper and eleven farm workers, who are to be passed along to Peter with this new tenancy. Doug Reid has limped on through the two decades that have decimated other farms. The Great Depression, which began in the USA in the 1930s, rolled its way around the world like a tidal wave to countries already depleted by war, culling Britain's world trade by half. Times were so bad that by the start of the Second World War in 1939, so many farmers had gone out of business that Britain was importing 70 per cent of its food.[2] When the war began, food became an easy target. Submarines sank the boats shipping it to Britain and blockaded the Atlantic Ocean as the Nazis tried to starve Britain into surrender. There was little option: the country had to find its way back to food production. *Ploughing on farms is as vital as arms!* The slogan worked. Six million acres of meadow grassland were ploughed up to grow cereals.[3] But by the time the war ended in 1945, three-quarters of the food consumed by the people of Britain was produced there. Some say this helped the Allies win the war.

Through all this Doug Reid hung on, scraping by, somehow avoiding the farm's requisition by the government during the war and seeing it into better days. Inside the farmhouse, though, little has changed since Victorian times. The primitive kitchen has a well for water and a hole in its roof big enough to see the sky through. The only heating comes from the fireplaces. The lavatory is outside on scrubland that cannot yet be called a garden.

But now it is 1959, and Michaelmas, and time for a change. Peter is two inches taller than Doug and his belly is flat and firm. A local journalist will later write, 'The tall debonair Peter Flindt could easily have been a film star if he hadn't taken to farming at such an early age.' It must have been the black moustache that did it.

This is not the only change. Farming is about to boom. Wartime farming subsidies and government grants have become peacetime policy. It is the beginning of advances in agricultural technology, machinery, seeds, pesticides and fertilizers that will result in astonishing efficiency and productivity, lifting millions out of starvation.[4] Those who invent the process by which artificial nitrogen is made – Fritz Haber and Carl Bosch – have long ago been awarded Nobel Prizes. Wartime chemicals designed to kill humans are diluted and sold as insecticides. Small meadows filled with hundreds of species of wildflowers and plants are no longer needed to feed the horses that once pulled the ploughs now they have been replaced by machines. Those meadows that survived the war are reseeded with crops or hardy ryegrass and artificially fertilized, which means that the wildflowers – which will thrive only in low-fertility soil – don't grow back. In the end, 97 per cent of all Britain's wildflower meadows are lost.[5] Many of the hedgerows that divide the fields, some dating back to the Bronze Age, are taken out to make room for crops. Wetlands are drained and rivers straightened. Trees are removed in their thousands: after all, farmers grow food, not oaks.

This year – the year that Peter arrives on the farm – is a tipping point of change, of growth, of revolution. This will become a year which will tell us about our future. The first satellite picture of Earth is transmitted from space. The Caspian tiger in Iran is killed. The first known human death from HIV occurs in the Congo. In the USA, the first commercial photocopier is launched. So is Barbie.

A few days later, on Saturday 3 October 1959, Peter returns to his new farm in his new Land Rover to start his new life. Next to him sits his wife. In the back, alongside her two siblings – for the brother who will one day work this farm has not yet been born – sits a six-year-old girl. Two decades later, this girl will become my mother.

*

On Peter's new farm, like farms all over the country, life revolves around food. After all, that's what a farm is for. The country only ended rationing five years earlier, nearly a decade after peace was declared. The low-slung fear of starvation is on everybody's mind long after wasting food stops being a criminal offence. Hunger is real. It is tangible and frightening and utterly possible. Fear of it lies within all those who knew it. For the rest of their lives they will treat food with care and respect. They will swill the milk bottle with water to get every last drop. They will cut ham when it has grown cold so as to get thinner slices. Food scraps will be fed to chickens, geese, pigs or whatever other animal will eat them. They will save the fat after cooking meat and the dripping will sit pale and slimy in a stoneware bowl, waiting to be reused. Those who want to cook with olive oil must buy it from the chemist. Avocados are foreign; holidays are not.

On Peter's farm, every kind of vegetable is grown in the garden, and fruit in a cage.[6] Raspberries, strawberries, black and red currants and gooseberries are eaten when they ripen, until the purchase of a freezer means they can be kept for winter too. Fruit trees are planted and grow into an orchard. Leftover pears and plums are stewed and frozen. The apples are laboriously laid out on newspaper in the cool of the cellar so that the air smells sweet and fermented, each fruit set down carefully to ensure a rotten one cannot touch its neighbour and spoil it. Scraps of meat are minced and turned into cottage pie for the coming days. In autumn, braces of pheasants hang on the back door and more in the cellar, waiting to be gutted and plucked. The entrails of a rabbit are regularly laid across the kitchen table as its flesh is made ready for a pie. Milk comes from a neighbouring dairy farmer, who appears every other morning with his churn and stamps muddy boots across the kitchen and into the scullery, where he pours fresh milk into a huge bowl. Afterwards the cream is skimmed off the top and two wooden pats brushed back and forth to turn it into butter. Eggs come from the chickens who scratch and peck amongst the orchard, while elsewhere in the country oversized barns begin to be built with cages for their cousins.

The farmhouse table hosts four meals a day, the large helmet of the village police officer when he calls for tea, neighbouring farmers, grain merchants, sales reps and the vicar. In a day it will see a cooked breakfast, a two-course lunch, tea with sandwiches and at least two home-made cakes before supper comes around again. There is enough food; more than enough. But they will still remember the hunger.

When he arrives in the farmyard in 1959, Peter's job, and therefore his purpose, and therefore his identity, is clear. The country – hell, the world – needs him to grow as much food as he can, in whatever way he can. His government will pay him to do it. His community will admire him for doing it. His mission is set.

He will produce beef and crops in the fields that sweep down into the valley from the farmhouse where he and his family live. Three-day-old calves will arrive in the calf pens and my mother will teach them to drink from a bucket of milk, their sucking reflexes pulling her small milk-covered fingers into their mouths. From January to May, eighty calves are sheltered from the winter in the calf pens and yard opposite the farmhouse, then let out into the fields when the spring warmth comes. They are fattened on the pasture around the farmhouse and sold for beef when they are two years old. Unlike the cows Peter once milked there are now too many for them to be given names. These cows have numbers instead.

In the decades after Peter arrives in Hinton Ampner, farming is regulation free. There is hardly a form to fill out or a permission to be granted. Hedges can be trimmed whenever they look untidy and not just when the birds nesting in them have fledged and flown, so my mother will collect their eggs on cycle rides, pricking one end with a needle and blowing out their contents. Whatever is unwanted is buried or burned or thrown away. An old pit is filled up with load after load of chalk after a chance encounter with a full lorry looking for a landfill site brings the two happily together. This buries not just the dell but all the rubbish thrown into it by the villagers who have, for decades, viewed it as their local rubbish tip. Keys are left where they are useful – in the ignition or in the door – and no one turns them who is not supposed to. There are men with clipboards, but they only appear to encourage Peter to grow more and more, for that is what is wanted. And so even a tenant farmer can have the sense that his position is secure: he is to grow food for the nation and he may do so in whatever way he pleases.

Farm life is free and unconstrained. It can also be hard and dangerous. Death is everywhere. On a neighbouring farm, the family's teenage son falls from a top loader onto the barn's concrete floor and is killed. Part of a farm worker's finger is lost in the tip of the combine and the children scramble in the straw to find it before packing it in ice and driving it and its owner to the hospital to be sewn back on. When someone sets fire to a barn and the asbestos sheets on the roof explode, everyone gathers around to watch the sparks fly. The Land Rover, with a top speed of forty miles per hour, has pockets filled not just with baler twine but enough rat poisoning-cyanide to kill a curious child. Unwanted stray farm kittens are rounded up, dropped into a cloth sack and drowned. Nests of baby field-mice are thrown into the fire and, in the spring, rooks' nests are shotgunned down with the fledglings inside them to stop them feeding on the growing crops. And, alongside everything, there remains the spectre of debt in a world where cash flows often do not keep pace with overdrafts and bills. A local farmer, driven mad by it, shoots dead his wife and son and daughter before taking his own life on the same day another of the children is due to visit them.

Harvest is the climax of the year's work. If the weather is bad, doors are slammed and swear words muttered as the children cower out of the way, lest they be blamed for the bad luck. But when tea-time comes each afternoon, rugs and jam sandwiches are spread out over the field stubble. The farm workers drink cold tea in glass bottles prepared early in the morning and the children use the hollow stems of the straw to drink hot tea from thermos flasks.

All four children are needed for the long harvest days – an expectation so culturally rooted that our school terms remain based around it, regardless of how few families now send their children to help in the fields. But even when the children are old enough to find work elsewhere – behind a bar or typing the letters of a man in a suit – they still choose to come back for summer work. When my mother turns seventeen, she is allocated to corn carting at harvest time. Sometimes she drives the cabless tractor wearing only a bandana and a bikini because it is the 1970s, and one must therefore never miss an opportunity to tan.

If the straw is not baled and stored until winter then it is then burned, along with the stubble. The children light a sheaf at one corner and, once it is aflame, run along the side of the field lighting the rest of the straw as they go. They dash away from the whirligigs that twist and speed across the field, their shape made visible by the burning straw in their vortex. The fire is thrilling and powerful and so hot it splinters the flints on the ground's surface. Peter is careful in his timing and method: it might cost him his tenancy if the fire gets out of control. There is a controlled break in the middle to draw the two fires towards one another, and keep the burn within a specified area. It doesn't always work. My aunt, employed on another farm in the summer holidays, finds a sudden change of wind has her running for her life, a wall of fire licking towards her. She escapes, although her arm hair and eyebrows do not.

After the fire dies, kept away from the hedgerows and trees by a ploughed margin, the land is eerie and quiet. Plumes of smoke waft up from the charred ground from flames not yet out. Whatever had been living in the field has fled or is dead. The birds are silent. It is black and apocalyptic and ready, now, for the plough to come, and begin the cycle all over again.

There are still men working on the farm although, as the years pass and they grow old, machines arrive to do their jobs instead. Still, Peter feels his responsibility to them. He must find employment for those who are left. At 7.55 a.m. each workday he walks up to the barn to see the men and give them their instructions, then returns for breakfast, the newspaper and paperwork. After elevenses – tea and biscuits in the farm office – he drives around the farm trying to catch the men not working, as they spend their time trying not to be caught not working. He returns just before lunch to hush anyone in the farmhouse to better listen to the weather forecast, which the farm, and therefore he, and therefore his family, must live by. His mood for the rest of the day will be determined by it. The afternoon brings other jobs around the farm or garden. Each Monday he drives into town to see his favourite teller, who will hand him cash for the men's pay. Each Friday afternoon he will hand his men the money in square brown envelopes. Unless it is harvest, everyone has the weekend off. And then, the following Monday, the week begins again.

In his farm diary Peter's words spell out an anthem
of progress. They are a song of change: a farming
revolution.

Shocking, threshing, carting, bagging,
morning milking, straw for thatching,
cutting thistles, rolling barley,
corn rick, straw rick,
Re-shocking fallen oats.

Drilling, harrowing, cultivating, ploughing,
all day 'tater-picking,
hedge laying, pulling bushes,
burning couch, trimming nettles,
Growmore Committee meeting.

Red clover, white clover, lucerne, sainfoin,
hoeing sugar beet,
blackberry picking,
afternoon burning bushes.

All hands rogueing wheat, All hands rogueing wheat, All
    hands rogueing wheat,
All hands rogueing wheat, All hands rogueing wheat, All
    hands rogueing wheat.
Sunday.
Rogueing corn, Rogueing corn, Rogueing corn, Rogueing
    corn, Rogueing corn,
Rogueing corn, Rogueing corn, Rogueing corn, Rogueing
    corn, Rogueing corn,
Rogueing corn, Rogueing corn.
Finish Rogueing Corn. Weather dull.

Drilling barley, creosoting, buckwheat,
Evening burning stubble,
Hedge trimming, subsoiling, topdressing winter wheat,
broadcasting red clover in barley. Yr 1 ley.
First load bulk fertilizer in.

Calf pens, sowing rape, spraying,
afternoon picking sloes.
Loading, stacking, discing, worming,
rotavating headland strips.
Feeding bullocks, rabbit shooting,
rogueing black-grass.
Combining wheat. Combining wheat. Combining
    wheat.

All day in drier doing Oil Seed Rape
All day in drier doing Oil Seed Rape
All day in drier doing Oil Seed Rape
All day in drier doing Oil Seed Rape.

Start silage,
fencing, drilling, bagging up,
ICI on beans, nitrogen on rape,
Plough rape stubble.
All day combining.
Afternoon to Alresford Show.

And always, at the start of nearly every single day: *round
farm*.

It is 1984 when Peter's youngest son, Charlie, becomes a
farmer. He is twenty-three. His first day as a farmer is

also his first real day working on the farm. This is not working on the farm like he did as a child: forking silage into the calf-pen feeders on winter weekends or helping with harvest in the summer holidays. It is a weekday, it is raining, and he is unstacking feed bags from the back of a trailer. He is alone. His father, Peter, is doing what he mostly does now – farming from the farm office in the farmhouse. The two remaining farmhands are elsewhere, maybe working, maybe trying not to get caught not working. Now it is just Charlie and the farm.

*Huh*, he thinks. *Well, I suppose this is it, then. I suppose I am a farmer now.*

His life as a farmer could have started even earlier had a teacher not stopped Peter pulling him out of school to begin it. Instead, Charlie went to university, and to New Zealand, and has come back to the farm with a degree in agricultural engineering and a girl called Hazel who will, later that year, become his wife.

But, in 1984, the country in which he is beginning his farming life is changing nearly as radically as it was over two decades earlier, when Peter began his. It too will show a shadow of the future. This is the year that the first Apple Macintosh computer goes on sale. Brenda Hodge becomes the last person to be sentenced to death in Australia. The United Kingdom agrees to transfer Hong Kong back to Chinese rule. A newly deregulated London Stock Exchange is on the brink of becoming the world's financial centre and, some will say, laying the foundation for a financial crisis which will devastate the lives of people all over the globe. This is the year that

George Orwell imagined when he wrote of a world caught up in endless war, historical negativism, widespread discrediting of experts, groupthink and omnipresent government surveillance. Some will later say that Orwell was only out by forty years or so.

In the years since Charlie has been away, farming has been living through a golden period. Just fifteen years before he became a farmer those who knew declared that 'a beautiful balance' had been struck in farming efficiency.[7] Then, in 1973, the UK joined the European Communities. The Common Agricultural Policy, of which the United Kingdom became a beneficiary, wanted to ensure that Europeans would never go hungry again. Chains of supply and demand across its member states were rationalized; agricultural productivity and technical progress promoted. A staple supply of food at modest prices was ushered in by the setting up of a common pricing system. Farmers in all EEC countries received the same price, fixed above the world market level, for the food that they produced. Levies ensured imported goods did not undercut the community's.

The more food farmers made, the more public money they got. And, oh, were they making it. Tonnes and tonnes of the stuff, enough to make mountains of food – mountains of grain, mountains of butter, mountains of beef – more than anyone or anything could eat. Farmers produced so much so fast that the first journal article to warn of growing levels of food waste created by the glut was published just twenty-five years after rationing was fully abolished.[8] There was so much for everyone: they were feeding the world! Except, that was, when politics and poverty meant that food still did not make it onto people's plates.

In each European member state there had been a huge increase in both the quality and quantity of food. Now when people wandered into ever-bigger supermarkets they found things they had only seen on holiday: Parma ham, olives, brie, chorizo, breadsticks – it was all there, whatever anyone wanted. Supermarkets, keen to keep an edge, switched from local producers to European imports in an effort to give delighted customers more choice. This bothered British farmers, but also it didn't, for they were guaranteed a price whatever they produced, and paid to produce as much of it as they could.

Now, in 1984 – the year that Charlie becomes a farmer – he will do so during a perfect summer. When Michaelmas comes and the leaves in the copse at the bottom of Englands field turn shades of gold, Britain's harvest produces the biggest yield of wheat in its farming history. Farmers all over the country fill their barns with record-breaking crops. Local people talk of the 'Barley Barons' who drive around in gigantic new tractors on the fields and gigantic new Range Rovers off them. That summer more than 170 million tonnes of grain is produced across Europe.[9] Charlie cannot know it then but, at every harvest for the next three years of his new farming life, the world will produce more wheat than it can eat.[10]

Farms start to make so much food that, four years later, measures are introduced to try to stabilize production, but with so much overproduction no one can stop farmers' incomes from falling. From the time that Charlie becomes a farmer his income will drop, year by year, at a rate sharp enough to soon be frightening.

Now it is 1991. Charlie turns thirty and he and Hazel finally take over the farm, although not yet the farmhouse. They do so in a very different world. When Peter was farming, the average household would spend a third of their annual income on food. Now they spend just over 10 per cent, a figure that will continue to fall.[11] After Singapore and the US, the UK will become the cheapest place to buy food in the whole world.[12]

This year will also become a tipping point in a decade of change, of growth, of revolution. Set-aside is about to become compulsory: farmers will be forced not to farm in an effort to control the surplus food being produced. For the first time, farmers' subsidy payments will relate to the size of their farm rather than how much food they are producing on it. A pilot scheme is introduced to enable farmers to earn extra public money, although the words 'farming' and 'food' are absent from its title. The Countryside Stewardship Scheme is focused entirely on landscapes and the creatures that live within them and what on earth, thinks Charlie, does that have to do with producing food?

Two years later he sets off to the farm accountants without Peter for the first time, driving to a city he rarely visits. The recently harvested fields he passes on the way are no longer apocalyptic black. Stubble burning has just been banned.

The government that rewarded Peter for all that production is starting to use new words. Radical reforms, says the European Commission, are needed to redress the problems of ever-increasing expenditure, declining farm incomes, the build-up and cost of storing surplus food and the damage to the environment caused by intensive farming methods. *The environment.* It is not a term Charlie has ever really used before. Soon it will be everywhere. It will belong to a whole bank of words he loathes so much he cannot say them without mockery. *The En-vire-ron-ment. Nay-chure. Gween.*

By the time the millennium Michaelmas comes, the renamed European Union is about to expand eastwards. The 'Agenda 2000' proposals warn that further changes must be made to farmers' subsidies because of the 'major income differences and other social distortions' of these new member countries, which mean food surpluses might grow even larger. There are proposals to cut guaranteed prices in various farming sectors. Subsidies higher than €100,000 per farm are to be capped, which mostly applies to the 'get big or get out' farms that everyone was encouraged to form. Farmers' incomes have fallen by over 60 per cent in the last six years.

On the coast, lorries and trucks line up in autumn rain for a blockade to protest against soaring fuel prices. On their radios a newsreader warns that 'The situation for the modern British farmer has probably never been so dire, and a further rise in the price of fuel could kill many of them off.'[13] If Charlie owned his farm, then what was happening would indeed probably have killed it off. If Charlie owned his farm, he probably would get out. But he doesn't, so he can't.

Now Charlie is thirty-nine with a wife and three young children, loans to repay and a not-insubstantial overdraft. He has an agronomist who advises him on what to grow and when, and what to spray on it and when. The drier barn is now a grain store, where corn waits for collection by a merchant who decides whether it is good enough for humans or just for animals, before selling it at the price they think is best. There are no more farm workers.

Charlie begins to write car reviews for magazines and pieces for *Farmers Weekly*. He is so good they give him a weekly column, and so popular he will still be writing it over two decades later. Hazel will get a part-time accounting job to earn extra income. Food comes not from the garden but from a Tesco delivery van. The kitchen table no longer hosts four meals a day with two different kinds of home-made cake, because who has the time for that?

Charlie now lives in a farmhouse that belongs to a farm that now belongs not to a squire on a hill but to the National Trust. It is to them that he must now pay his rent. In return, the National Trust becomes one of a growing number of official bodies who control what he does and how he does it, for whom he must fill out forms and comply with inspections and – oh, farming is so very different from how it was.

On the walls of the farm office where Charlie now sits hang silver shields for farming prizes with Peter's name engraved on them.

'Bloody good farmer, your grandfather,' people tell me. 'Bloody difficult land, that.'

The farmer who almost never drove a tractor was certainly good at winning prizes. His son, who spends all day on one, doesn't win any. But then, farming means something different now. For this is the beginning of a new time. It is a tipping point in a decade of change, of growth, of revolution.

If someone on the street were asked to describe a farmer today, they might get close to Charlie. He is sixty years old – the average age of a British farmer – and big in every sense: body, voice, opinions, heart. He is a UKIP-voting, BBC-loathing, Facebook-posting, Brexit-supporting, climate-change sceptic in a checked shirt and wellingtons with a supply of one-liners. A lifetime of shooting and drier work has left him with almost no hearing in his left ear so that he must sit with people on his right and angle his face towards them with an expression of incredulity, which always makes me wonder if he hasn't heard me or just thinks what I have said is profoundly stupid. Both are equally possible.

He is plagued by a medley of illnesses after a lifetime of physical work – bad back, bad shoulder, bad gut, bad sleep – but is still a warrior of a man, the grey of his once-red hair and beard failing to lessen the force with which he fights his battles. His year is marked by them.

In autumn, when the farming year begins, so do the fights. The fights with poachers who rag their cars across newly drilled fields as their dogs rip off the heads of hares. The battles with rural thieves who now bring a selection of tools with them, the better to make their way past chains and padlocks. The battles with a rolling succession of ramblers, horse riders and dog walkers, who often seem surprised to be at the end of his wrath and claim not to have known there was a crop in the field they have just stomped across. The arguments with villagers who begin to use a small, fallowed field adjacent to their houses as their own after decades of it being left idle. He settles that battle by ploughing it up.

Then there are the smaller wars which concern him throughout the year. The BBC (*#Defund the BBC*), *Countryfile* (*utter bollocks*), *Farming Today* (*mostly bollocks*), George Monbiot (*agri-woke catastrophist*), Chris Packham (*same*), Greta Thunberg (*Swedish doom goblin*), vegetarians (*hypocrites*), local villagers (*white settlers*), land agents (*chinless ya-yas from up country*), government agents (*miserable squirts with a 2.2 in Environmental Management with Media Studies from the University of Central Rutland*), The Met Office (*muppets*), The National Farmers' Union (*No Flipping Use*), *Lambing Live* (*this is a bloody working farm not a TV show*), the new landlords of his local pub (*hardcore Remainers*), use of 'eco-', 'enviro-', 'green', 'natural' or 'sustainable' (contempt demonstrated by use of single quotation marks, typed or finger-indicated), any products containing said words (*as effective as semolina*), Natural England (*one of the many remote and mysterious unelected bodies that rule* *our lives*), The National Trust (*all 'can't, shouldn't, mustn't'*), Right to Roam (*what about the Right to Privacy or the Right to Earn a Living*), Regenerative farming (*ostentatiously claiming to be working with nature will not produce everything you'll ever need*),[14] non-farming critics of farming (*wouldn't know a mouldboard if it smacked them on the head*) and National Trust walkers, around whom he must plan his monthly spraying schedule so as to avoid their fury at him for doing what is, as he sees it, his bloody job.

But behind all the anger is this. Oh, how it sticks in the craw to be told he is destroying the world for doing what he has been asked to do. Oh, how galling it is when other industries that make products so much less necessary than food escape the same condemnation or, worse, join in. Oh, how it rankles to be lectured by hypocrites. To be told off for fossil-fuel-farming by those who take holidays on planes to places he never gets to go to. To be told he is killing insects and destroying wildflowers by people who live in houses built on top of farmland. To be thought cruel for sending animals he has cared for to slaughter when their life is no more or less valuable than the tiny creatures killed in the production of vegetables, or cotton, or bread. To be judged by city folk so entirely uneducated about farming that a group of walkers once ran away from his combine harvester in terror.

He has a point.

The memories of my grandparents' farm sometimes come more easily than those of the house we grew up in. Maybe this is because it offered all a child could want:

freedom, adventure, animals, cake. Or maybe there is a certain kind of alchemy that takes place when a child is allowed to explore the natural world unwatched.

I sit in our Suffolk cottage, with the land outside my window turning from summer to autumn, and type the address of my grandparents' farm – hundreds of miles away from where I am now – into Google Earth. I am an electronic snooper; a virtual spy. I zoom in on the fields, hills, woods, farmhouse and find my childhood self with the cursor arrow.

I hover virtually above the orchard where my sisters and I turned the hen-coop into the headquarters of our secret club (members: me, my sister, my other sister). I find where the two geese were kept – Hissing Sid and Selina – who would chase us up the garden with the ferocity of two Rottweilers. I find the tree in which I built a house of wooden crates tied on with baler twine, and the barn that once housed the calves in winter, where I would spend hours trying to make pets of them. The hay barn where my sisters and I made dens of rect-angular bales while dodging mouse droppings. The old oak tree where my friend fell and broke her arm. The barn whose walls I spray-painted with pictures for my eighteenth-birthday party, beneath beams painted with flowers by my mother for her sixteenth. The new barn, which my sister dressed in white drapes so she could pretend it was a marquee for her wedding. The grain-barn whose concrete wall bears my mother's scratched signature from 1965, and where I would go to find my grandfather at harvest time – dusty in a brown shirt, stained dark under his arms and on the back – so he could make me a small patty by pressing kernels into a hollow disc, then pulling down a handle to crush it into a corn cake for me to eat.

I find where my grandmother coaxed a garden into life through hours spent on her hands and knees in the dirt. I find the place where she grew so many daffodils that, when I lay down amongst them, it felt like I was drowning in a yellow sea. I find the garden slope where I would sit with her and shell peas into a stoneware bowl, and watch swifts fly in the blue above us.

I find the farmhouse where I once slept in a bedroom whose window looked out onto a copse at the bottom of a field called Englands, whose leaves would burn yellow and red each autumn. In this house my grandmother taught me how to pluck pheasants and make bread, rock cakes, shortbread, jam tarts, fruit cake, butter, batter and harvest teas as I stood on a chair and tried not to catch my hair in the sticky yellow fly strips that hung from the ceiling. I would watch her hands work and memorize the calloused finger-pads and swollen knuckles and skin cracked from being so often in water, which the Swiss hand cream by the sink never seemed able to soften. Once, when I was a teenager, I noticed that my grandmother had passed her hands on to my mother who, in turn, had passed them on to me. Now I look at my hands on the keyboard and see how my work has made them as soft as my grandmother's had made hers rough.

I wonder whether I loved my grandparents not so much for who they were, but for what they were. They were the farm. They were the outdoors and animals and soil. They were food and fruit and flowers and harvest. They were a hard-working life; a life that seemed to have both a clear purpose and a clear reward.

My grandfather died on the farm almost exactly fifty-two years after he first arrived. He died when the leaves of the copse were lit up with their autumn colours and the day after he chopped a winter's supply of wood for my grandmother, as though he knew death was coming for him. He is buried in the graveyard of the stone church where my parents, my aunt and my sister were married. When my grandmother died, two years afterwards, her ashes did not join him. They still sit in an urn in the farm office. One day, when no one official is around to ask him what he's doing, Charlie will scatter them in the fields of the farm, for it was here, mixed into the earth of the place that she so loved, that my grandmother wanted to take her eternal rest.

Now, like all the others who came before them, my grandparents have become a layer in the farm's history, a part of its patina. They have shaped it, for better or for worse. Even if the marks they made are extinguished by whomever or whatever comes afterwards, and even if their names and deeds are forgotten, in death they now permanently belong to the land that once temporarily belonged to them.

The house my parents now live in is the sized-down version that has little of my childhood in it. So it is only after I have left it behind and begin to drive towards the farm that the familiar lanes and views mean my child-self slips under the skin of my adult one. It confuses me. It always does. I find the scene outside my car windows so full of memories it feels almost wrong to now be driving through it with Ben. He belongs in my new life and not the one I have left behind.

We are driving through a storm to meet my uncle Charlie, which is fitting, for he is a storm of a man: an immense bluster, full of howl. By the time I have reached the pub my stomach is a twist of nerves. I park the car and try to wrestle adult confidence out of my teenage ghost.

Ben and I arrive before Charlie does. We push our bodies through the wind to the front door. It opens into a tiny lobby with a choice of entrances on either side. It is a test. To the left is the locals' bar, a small room where farmers sit to loudly complain and agree on all the things they are right about and fall into silence if the door opens to a trespasser. The right-hand door opens into a space with wooden tables, lit fires, small windows set low. People at the tables glance up as I walk past them to the bar. I have worn something anonymous – blue jumper, jeans, trainers. They are clothes chosen to hide the fact that I am now a foreigner here.

I order drinks and am rebuked by the man behind the bar for saying that ice in my Coke would be 'amazing'. He is right, but the pedantry still stings. I remember this is somewhere I must keep my language modest.

Through the low window I see Charlie arrive, bending his bulk against the storm. Inside, he goes to the bar and is handed a thank-you card by the publican. Inside is a ten-pound note – a gift from a fan – to have a round on him. He is famous in this world of farming. If ever I tell a farmer I am his niece they look me up and down and grin: 'Charlie Flindt's your *uncle*?' as though they had secretly suspected the man was mythical.

We are careful with each other. We are family. He goes gently, even though I know that, in so many ways, I represent everything he loathes. I am a city-dwelling liberal who spent the referendum campaign pushing 'Remain' leaflets through doors. I've supped on the Kool Aid. I am a 'soil catastrophist'. I have too readily accepted arguments that find his degree and lifetime of farming experience 'trumped by reading the *Guardian*, watching *Countryfile* and harvesting echo-chamber opinions at SW1 twatterati dinner parties'. And now, on top of it all, I've decided I fancy having a go at farming.

We part from the pub as friends but it does not stay this way. Over the following months we argue in the modern way: by keyboard fight. He is the only person who communicates with me on Facebook Messenger. I begin to dread its ping. We both flare; go silent; sulk; cry alone. It teaches me that we are more alike than I knew: quick to flame, quick to cool and a little too inclined to think we're right.

Charlie has been intensively farming for forty years and he cannot see any reason why he would want to stop now. He uses all the farming tools he can, both chemical and mechanical, to grow as much as he is able, for this is still his job. As far as he is concerned, given our ever-growing population, being a farmer still carries with it a moral duty to produce as much food as possible. There are still people going hungry and many who survive only because food is cheap.

Charlie is sure that the soil on his farm is exactly where it was when my grandfather first arrived to farm it. He is sure that nothing he has done has harmed it, eroded it, depleted it, sterilized it, or otherwise crop yields would not still be rising. He thinks blaming cows for methane emissions is insane (*heard of the methane cycle?*),[15] and capturing carbon is nonsense (*ditto the carbon cycle*).[16] Hedges should be flailed regularly and any creatures living in them will be undisturbed because only the soft regrowth is cut. Sightings of owls, hares and farmland birds are proof enough for him that wildlife loss has at best been greatly exaggerated and at worst is a phenomenon in other places to which his methods have not contributed. Bites, stings and wrecked crops are all the proof he needs that there are no real issues with insect decline. If there are fewer insects splattered on our windscreens it is not because of insecticides but because our cars are more aerodynamic. It's the return of raptors that should get the blame for songbird losses, not farmers. When the lanes are congested with excited twitchers around his farm because a red-backed shrike had been spotted nearby, it is not proof of the bird's rarity but one in the eye for the doomsayers.

But then, there is this. He protects tiny leverets he finds when combining, shepherding them to the safety of the hedge away from circling buzzards. He is haunted when a rabbit goes under his tractor wheel; distraught when the farm cat goes missing. When a seagull is trapped by a clod of ploughed earth he goes to some trouble to rescue it. He takes joy in the sound of birds singing in the spring and is sentimental about the farm and its history. In short, he is more than the sum of his parts. Well, aren't we all?

Charlie has, over his farming life, paid nearly a million and a half pounds in rent to farm this land and yet still he owns none of it. All he has for a lifetime's work is his pension, the machinery he has paid off and a part share in a grain store. And he has his legacy. But now that the agenda he has railed against for so long has become The Agenda, it might just do away with that as well.

When my sisters and I began to grow up and pull away to city lights, Charlie would tease us for our townie ways. Shiny white trainers, cropped tops, artfully ripped jeans. On a blackboard he once chalked up a mock menu of all the ways he could cook and serve our pet hamsters and guinea pigs, complete with drawings. In a way the jokes always had the same heart to them: look at you, moving ever closer to a place of foolishness. A place where practicality is superseded by vanity. Where animals are pets, not food or tools. Where life is not real. Because real life was on the farm, in a place where work was hard and physical and purposeful and more honest than life behind a computer. It didn't matter that the internet connection at the farm was lousy because what did farming have to do with the internet? Banking could be done in person or by cheque or phone and the tax worked out by hand in a book. But the teasing was bitter-sweet because ultimately, as the years passed and the overwhelming majority no longer lived his life, it must have begun to seem like he was living in the unreal world, and that the world behind a computer screen was becoming the real one.

To be eligible for public money in the future Charlie will have to change the way he farms to fit in with all the 'eco-jargon' that makes him shudder: sustainable farming, nature-friendly farming, biodynamic farming, regenerative farming, agri-ecological farming, environmental farming. Charlie does not want to farm this way because, to him, it is blindingly obvious that the world cannot and will not be fed with lower yields and fewer crops. His job is the same job it was in 1984: to feed the world.

Now, though, he is sixty and the world is different. Now the world throws away 2.5 billion tonnes of food every year, eating just 40 per cent of all the food produced.[17] Now a third of all fruit and vegetables never make it onto a supermarket shelf because they are the wrong shape or colour, or because the farmer produced more than the supermarket ordered, or because the supermarket simply changed its mind.[18] Now the world makes enough food to feed three billion more people than actually exist, even if it doesn't always get to those it should.[19]

Farming today is about to mean something entirely different. For this is the beginning of a new time. It is a tipping point in a decade of change, of growth, of revolution. Charlie is the last person alive who has worked every part of this farm. His children do not want it, or this life. Why would they, having borne witness to the struggle of the last twenty-five years? Besides, they can get better-paid jobs in a city in front of a computer screen, working in what has become the real world. So I wonder what will become of this farm? What will happen to this place I somehow imagined would always, at least in theory, be in my life?

On the sole of the farm's boot-shaped boundary there was once another farm. It had a small house with a cottage garden, a yard, some barns and a dew pond for livestock to drink from. When Peter, my grandfather, first arrived in the farmhouse up the road, the old farm's buildings were all still there. There, but empty. The year after he arrived they were bulldozed to the ground. Parts of the old farmhouse were then used in one of my grandfather's new barns. This old farm, and all the families who had lived and worked there, were forgotten. They left behind their ghosts, though. The garden soil always grows a better crop than elsewhere in the field that consumed it. And sometimes, when the weather is right, the ground will dip where the dew pond once was, showing a shadow of the past.

I don't know what will happen to Charlie's farm when he no longer has the energy, inclination or funds to farm it. The National Trust, England's second largest land-owner, has announced plans to plant 20 million trees in the next decade following a £4 million donation from a global bank.[20] Its director-general has made it clear it intends to focus this on its farmland. Trees, says the

National Trust, 'are our best natural defence against the climate crisis'. An increasing number of the Trust's tenanted farms are being taken in hand, the fields rewilded and their farmhouses turned into holiday lets.[21]

If this is the fate that awaits Charlie's farm, which has grown food in its fields since the 1600s, it seems unlikely that anyone will remember or know how to see its farming ghosts. There is something else. If the farm stops being a farm it is not just Charlie's legacy that will be lost, or my grandparents', or Doug Reid's. This house, this place, this farm, is also part of my history. And, if it goes, part of me will be lost with it too.

*The biggest obstacle is what's between your two ears.*

Tim, farmer

# 3. Year One – Slow Walking

I am determined to learn about this new life I accidentally find myself in, and so I start to read nature
books – bestselling ones, not academic ones – which
then just tell me how much I do not know. The authors'
paragraphs are littered with names of flora and fauna,
insects and birds, and I find myself skimming over sentences made up of words that make no pictures in my
mind. This world of nature they describe seems a rever-
ent and unknowable one. Sometimes it feels like it exists
in a perspex box that the authors are studying, carefully
turning it around in white-gloved hands for examination. I find myself looking for the nature I am living in,
which bites and stings, burns, scratches, unbalances,
whips and bends with wind. With the memory of birth
so vivid in my mind, I know that nature is frightening,
uncontrollable, animalistic, unsentimental. That it can
hurt as often as it can heal.

I wonder often how the writers can go for such long
walks without other people needing them. How are they
free to stride into nature for days, weeks, years at a time
so they might describe it to me? Maybe they have chosen
a child-free life, the better to be wolfish. Or maybe there
is a silent supporter left at home with weaning spoons
and potty-training pants and a Calpol syringe lost down

the crease of the sofa. Nietzsche wrote: 'Only those thoughts which come from walking have any value.' But then I don't suppose he did it with a baby on his back and a toddler by his side, trying to separate his valuable thoughts from the babbled commentary of a newly discovered world.

I cannot take long walks into nature. My children tether me both physically and practically, and sometimes mentally too. My smallest son Aubrey will take his first steps on uneven ground. Later, I will wonder whether it is this need to balance well that makes him such a solid and grounded child. Wilfred is two years older and entirely different. Even so, he has taken to this new life as though his old one was not spent playing with nitrous oxide canisters in grey city playgrounds. Up until now his journeys have been spent strapped into car seats or buggies pushed along pavements at the level of car exhaust pipes. Now he is unstrapped, I cannot keep him in. He escapes almost daily over the fields. Once he is returned, barefoot and eating an apple, by a neighbour who spotted him wandering up the road as I was upstairs changing the baby. I am both terrified by and in awe of his new independence. He is uncontainable, almost always missing some of his clothes and mostly unwilling to pee inside. He begins to learn country tricks. He rubs dock leaves on nettle stings to stop the pain, even though I cannot tell him why it works. He plucks the white flowers of dead nettles and sucks out their honey centre. He befriends huge spiders, woodlice, moths and grasshoppers, and lets them run over his bare torso. I watch him watching this new world around him. As I look at it through his eyes, I start to see it differently.

He begins to ask me the names of plants and trees and birds and creatures. I am ashamed that, even though my grandmother would have once taught them to me, I have forgotten them. I hadn't needed to remember. I begin to learn them to teach him and in doing so teach myself. I download an app on my phone that can identify plants and use it to name the wildflowers as they start to appear. The more I understand about the natural world, the more I notice it. The more I learn, the more I hear it. Sometimes it feels as though I have been living a half-life. It is like deep-sea diving: like going into the water when you have only known its surface and discovering there is a whole world you had never noticed living underneath.

But my ways of experiencing this countryside we are living in are different from the ones in the books I read. My children are not, I learn, good at a stake-out. I cannot sit still and wait for a fox or vole or field-mouse to patter along without one or both boys beginning to become cold, or hot, or hungry, or thirsty, or bored — so very bored. Instead, our walks are speed-limited and unsilent. We are bound to the pasture fields at the back of the cottage, for that is the distance that little legs can walk. So instead I begin to learn every inch of them. I learn what they look like at dawn and mid-afternoon and in the summer dusk or in the cold indigo of twilight. I learn to notice new leaves or blossoms where there had been none. I see how their shapes change with the seasons.

Because I experience it this way, nature becomes kaleidoscopic. I see and feel it not just through my own prism but through my children's too. After one of Wilfred's escapes I find him crouched motionless, wearing only pants and boots, watching ants streaming in and out of an enormous earth mound. When he sees me he cries out – 'Looooook, Mama; look what I found!' – and I stop myself telling him off for running away again or pulling him home. I squat next to him with the baby on my knee. We watch the ants' procession. I see how they avoid bumping into one another, despite their single-minded speed, and how occasionally one will be carrying a portion of leaf bigger than its entire body. I think of how I haven't watched ants this closely since I too was a child. It is a kind of meditation.

One August afternoon I am walking with Aubrey strapped to my front and Wilfred's small, fat hand in mine when he spots a raptor on the ground. It is hidden by the long grass near the white willow tree in the lower field, closer to his eye-line than mine. It is streaked brown, yellow beaked, and camouflaged against the late summer grasses. I wonder if it is a buzzard, one of the pair I see and hear wheeling over the copse. Sometimes their mewling cries sound so much like a new-born baby that I catch myself looking up at the cottage bedroom window, in case there is a child there I had forgotten about. The bird fixes its black eyes on us and I realize I have never seen one so close before. I take a photograph on my phone and later decide it must be a kestrel, smaller than a buzzard with cream-coloured chest and legs. To me, raptors still seem a little mysterious. They are not the birds of my childhood. The pesticide dichloro-diphenyl-trichloroethane (DDT) thinned their eggs so that they broke before the chicks could hatch, but even after they were given legal protection in 1954 most raptors were still rare. DDT was not banned in the United Kingdom until 1986. Afterwards, raptor numbers soared. Some blame them for the crash in small-bird populations, but when I look it up I find that every study undertaken has found the opposite.[1] When I show Ben the kestrel's photo later on I realize that if Wilfred had not been with me, I probably would never have noticed it.

One morning, later the same month, I am up early with Aubrey, at dawn. We sit together in the lean-to which overlooks the garden and meadow-field and watch the sun come up. It is so beautiful that I take him outside in our nightclothes to greet it. The field is dazzling silver. At first I think it must be dew, but when I walk into the long grasses I realize the shimmering is thousands of tiny, perfect cobwebs. They are lit up by the morning summer sun, swaying and sparkling like a West End show. They cover the whole meadow: a secret exhibition created in just one night. I crouch down to better see the strands flowing from one tall stem of grass to another, joining the field together in a delicate miracle of engineering and art. I later learn that, at this time of year, spiders try to catch as many insects as possible to feed themselves up before they mate. After mating, the female will often store the sperm until she is ready to produce eggs. She will construct an egg sac from strong silk, then lay her eggs inside, fertilizing them as they emerge. When the spiderlings hatch in spring they will climb to the tallest stalk they can find, raise their abdomen and release threads of silk from their spinnerets, which hardens as it hits the air to become threads of gossamer which catch the wind, carrying the spiderling away.

When Ben and Wilfred are up, I pull them outside to the field to see the webs, but in the daytime August sun they have all become invisible. Months later, when I see wisps of silver gossamer drifting through the air, I wonder if they have been made by the spiderlings of the spiders who built the webs that morning.

I notice how both children seem entirely untroubled by dead things – of which there are many – and entirely accepting of death: existentially, anyway. They are fascinated by skeletons of birds and mammals and will poke and pick over them with sticks if I'm watching and fingers if I'm not. They watch Ben break the neck of a myxomatosis-diseased rabbit with nothing but curiosity. It is suffering from a virus intentionally introduced in the 1950s to control the wild rabbit population, which was destroying great swathes of land. Wilfred stares at the milky white of the infected rabbit's eye and seems to accept that hastening its death is a kindness. Still, it's hard when there are so many of them running away from us not to think of rabbits as pests, rather than a food source for predators. Some of these predators were brought to the edge of extinction when rabbit numbers fell. My grandfather used to treat them as his morning sport, shooting them out of his bedroom window before breakfast. Ben once does the same. Wilfred learns how to skin and gut it for eating, unbothered by the slippery, bloody organs all over the kitchen worktop. It is me, not him, who turns away.

I see my children wielding sticks or throwing them like javelins or stalking wood pigeons and think how fundamental their predatory instinct seems.[2] I wonder when I stopped thinking of myself that way. I tell them of the purpose of the cows and sheep that graze in the fields around them because I refuse to pretend otherwise. I tell them that it is respectful to know. They look at me with earnest faces, trying to understand. But then they have seen the natural cycle play out beyond their window: a hunt, a kill, an eating. They have borne witness to the bloody aftermath of death: half-eaten flesh, skeleton bones, a pigeon wing or head when the sparrowhawk has come out hunting. They are learning that every being is a link in a chain, all connected one to the other.

For supper I serve up sausages, cottage pie, bacon, drumsticks, mince, and wonder – now it is living outside my window – when we started to prettify what we call our meat. A long time ago, it turns out. Our 'French' vocabulary, with its Latin origin, was imported to Britain following the Norman Conquest in 1066. While English people continued to use their Anglo-Saxon language they also became influenced by the words of the ruling elite. Meat began to be described in doublets. When grown and slaughtered by Anglo-Saxon peasants it would be 'oxen'. When it made its way onto the plate of Anglo-Norman nobility it became 'beef'. It was the same with calf and veal, sheep and mutton,[3] deer and venison, pig and pork.

I start to wonder whether it is not just what we call our food that lets us separate our eating of it from the truth of what it is, although that helps. Cities, generally, help us to distance ourselves from all sorts of truths and responsibilities. Drop a can or cigarette butt or crisp packet in the city and someone else will soon come along and sweep it up for you. Here, in the countryside, it will stay until it is buried by leaves. Leave a gate open in the city and no one will be able to tell whether it was you or one of the other hundreds of people passing, and what does it matter anyway. Here it can mean cows running up the high street and sheep grazing on someone's front garden. Pass a stranger in the city and you can look straight through them, for they are the hundredth stranger you've seen that day and you will probably never see them again. Pass a stranger in the countryside and you will be unlikely to cross paths with anything less than 'Hello'.

I begin to wonder whether our solipsistic cities have promised us that we need not really be responsible for anything but our own wants and desires. Many of us have utter liberty to be who we want to be, which means that, for a lot of the time, we can behave how we want, dress how we want, party how we want, eat what we want, listen to what we want, read what we want, say what we want (although mostly to people who agree with us, unless it is behind a Twitter avatar). Maybe this is right. Maybe this is progress; maybe this is freedom. But still, the consequences of our choices are often kept hidden away from us as though we are children. God is dead; so is society. We are told that the only person we need to be true to is our own authentic self, but when we look within ourselves for answers to life's biggest questions we are shy to say when we find emptiness instead of insight. In all this there is room only for life, and none for death. Death is kept away from us: hospitalized, sterilized and shrink-wrapped. Maybe that is one of the reasons we try so hard to ignore it, to escape it: the only event certain to befall us all. Maybe that is why it can be so shocking when we must finally confront it.

Summer fades into autumn. At the far left corner of the meadow behind the cottage, near the copse where the barn owl flew, are blackberry bushes that leave my sons with hands and mouths stained blue-black. In the garden the wind has blown the leaves into a riot. They prostrate themselves over the grass in outrageous shades of red and yellow and brown. There are so many, I pile up a bed of them on which my children bounce.

The colour of the leaves is not the only thing aflame. On a bright September morning, Ben calls me – 'Turn on the telly – all our stuff's on fire' – and I start to feel like maybe I am living in a parable.

The BBC news website shows a picture of the warehouse where we had stored our belongings when we left London. The whole building is ablaze, an immense pyre rising thirty metres into the London sky. The smoke that blooms around it is tinted red like it has been lit up by stadium lights. 'Fire Crews Tackle White Hart Lane Blaze' shouts the headline. The fire lasts for three days.

Afterwards, I ask whether any of it is salvageable. Our brass bed? Cutlery? My bike? Gone, all gone, the loss adjusters tell us. No adjustment to our loss, then. It is total.

I must start itemizing everything I own – everything I owned – to make our insurance claim. I wonder whether losing almost all our possessions will undo me. Whether it will leave me undefined, somehow, wondering who – without the collection of myself – I am any more. I look at the photographs used by the estate agent to advertise our old rented home to try and remember what the life we once had looked like. It is a strange experience, picking over the bones of something that is starting to feel like someone else's life, a bit like looking at a photograph of yourself in a place you do not remember visiting.

Some of the items are easy.

47.  15 pairs of black Marks and Spencer tights –
     £5 each – bought between approx. 2005–2014

75. Three-seater sofa – £1,000
102. Television – 2008 model

Some less so.

7. Foetal sonograms – £?
54. Framed wedding announcement and
    wedding order of service – £10?
279. Great-Grandmother's wooden sewing
     box – £30?

I should not be surprised by the burning. After all, this is the month of fire. The Romans believed September was overseen by the fire god, Vulcan. Once, the fields outside my windows would have been blazing too. This is the season when farmers burned the stubble left after harvest to prepare the soil for a new crop. The practice was banned in the UK in 1993 for being truly pollutive to the atmosphere and blackening the homes and lungs of those who lived near farmland. The microbes within the soil were no longer seasonally torched, and the animals and birds who had made their nests in and around the fields no longer, like the ladybird in the nursery rhyme, found their homes on fire and their children gone.

Farmers set their fields on fire believing it broke down all the dead matter left from harvest and released the nutrients so they were available to the soil once more, clearing the ground of weeds and disease at the same time. In England some controlled fires are still used to restore and protect rare heathland and moorland eco-systems. In other countries, such as America and

Australia, fires are still used to clear dead wood and let the sun in. The destruction caused by fire allows new growth to start.

Is this what is happening, I think, as I spend my evenings looking at pages and pages of Excel spreadsheets, itemizing my old life in gridded squares. Did my life need to be broken down and cleared out so that the light could fall on it again? Was all this stuff, which I thought made me who I was, really just dead wood? I look outside the cottage windows into blackness and notice how the nights are falling earlier and earlier. I think, *Go on then, what else have you got for us?* – which is a reckless come-on to a universe that is beginning to feel malevolent. If I believed in signs – which maybe I do – this feels like the clearest one yet.

It snows that winter. It is the first time the children have seen it. The cottage has no central heating or insulation and so we are aware of the winter in a completely new way. This time the season cannot be kept outdoors. It follows us inside.

Winter now means red hands and noses and seeing our breath both indoors and out. I learn we must wear socks and slippers and jumpers indoors, and dress and undress as fast as possible at morning and night. I am aware of the chemical reactions that take place between heat and cold, which see the inside of the windows frosted in the morning and running with condensation by lunchtime. A wall by Aubrey's cot becomes spotted with black mould from the damp and makes me worry for his lungs. I am learning about the work and preparation it takes to make the cottage warm from just a fireplace, a wood-burner and an oil-powered Aga, which we huddle around as if it were our mother.

One day, we run out of wood. We are still too complacent, unused to the need to prepare ourselves for winter, and the woodshed outside is empty. The copse two fields away is full of dead wood and fallen branches, but I now know these must be cut and stacked and dried out for months and months before they will burn safely. Ben is away, so I dress the children in layers of jumpers, scarves and hats and leave them with their babysitter, who has come for the day so I can work. When I step outside, the cold hits my face so sharply I feel my skin contract over my bones. The car is buried in snow and so I walk to the village instead, carefully, slipping along the ungritted road.

The village looks like a postcard. The pastel-coloured houses with wonky windows that line the steeply sloping pavement are topped with white. Some have front doors and rooms which are so small they look like playhouses, for they were built for a generation of people who did not have as much plenty as ours and who would have thought me a giant. Some children are using bin liners to sledge the pavement from its top to its bottom. Their yells are the only sound: everything else is muted by snow.

There are no cars on the road today and so I walk carefully up its middle. This road makes up the whole of the village. On either side of it are places that are finally beginning to feel familiar. There is the greengrocer, with faded serving suggestions on the labels (*Avocados: slice in half and fill with prawns and mayonnaise with a squeeze of lemon*) and where purchases are totted up with a stubby pencil on paper and paid for in cash. There is the news-agent, which trades penny sweets from hefty screw-top jars, hand-knitted baby cardigans, toy cars, magazines and local knowledge. Opposite is the antiques shop, which also sells birthday cards, candles and soap. Fur-ther along is the butcher, who pins up a map showing where the meat has come from. At the end is the Co-op, with a locals aisle and a noticeboard advertising jobs and machinery and where no one minds queuing to pay because it means you get to talk. Finally, there is the hardware store, Abbotts, a Tardis that sells everything from wellingtons to paint to door handles to bird seed to baking trays. Ben once thought he had found an appli-ance they would not stock only for the owner, James, to reach below the counter and pull up two boxes: 'Large or small?' Now, I make my way inside and buy neatly cut logs held together in a string net.

As I walk back with the wood over my shoulder, I look at the fields either side of me, sloping up and away from the road. I think of a new farming word I have learned from the reading I have been doing: *vernalized*. The seeds of some plants, such as winter wheat, need to go through a period of intense cold to be able to flower, which, for wheat, means the growing of grains. Vernal-ization, which comes from the Latin *vernus*, 'of the spring', can happen only if the seed is exposed to the bitter cold of winter.

I think how winter is not a deadness. It is not an absence, either. I am learning that this time is a necessary part of the life cycle. It is in winter that the frost creeps its way around ploughed soil, freezing and swelling the water within the soil particles and cracking it into a fine tilth ready for sowing. It is winter that brings other seeds to life, the frost softening their hard seed coat so that the seedlings might germinate and grow, searching for sun and nutrients. But maybe it is not just plants that need to undergo a wintering before they are able to grow or flower. Maybe we must too. I am starting to wonder if this accidental time away from everything I had thought was important – this time I had not planned, sought or wanted – might not be an absence of life at all. That it is more necessary than I had known it would be. That, maybe, it is me who is being vernalized.

Later, I will learn the same lesson from a farmer, Oliver. As I was trying to work out whether the blueprint of a farming family would be enough to learn the land, he was on the other side of the country trying to do the same. His story showed me that, so often, we end up finding a route back to our past. Ollie will teach me about farming and food systems and ways to regenerate the land, but he will also teach me that when life seems frozen, maybe, in fact, it is actually just a preparation, a readying of all that is needed for us to begin to grow.

*Part of it is a practical decision, but there is*
*definitely a spiritual angle.*
*It's a very nice, calming, peaceful way to farm when you*
*are working with nature and trying to do things which*
*have a positive effect.*

Ollie, farmer

# 4. Ollie

In the summer of 2018, a drought burns through England. It is one of the hottest for over a hundred years. The driest since records began. Its heat parches the soil and makes the past visible. Deep in the earth, buildings are buried. The soil above their walls dries out faster than the soil around them so that a skeleton of what is hidden underneath is scorched onto the ground above. They are called crop marks, a dull name for something extraordinary. An undiscovered Roman road is burned from the earth in Wales, cutting across fields and right through the middle of a farmhouse, straight and blonde and long. Cambridge students lounge in the sun alongside the outline of a Second World War air-raid shelter. Historians are as giddy as historians get at the discovery of previously unknown Neolithic ceremonial monuments. Two Iron Age settlements. A Roman farm. Countless former castles. Funerary monuments, ditches, moats. Nearly two hundred ancient sites reveal themselves under the burning sun. Some of them date back to 3000 BC: a time so far away that, if you stop to think about it, the idea is hard to put your hands around. History is changed; accepted facts are altered. The soil, it turns out, keeps secrets.

When the rain falls and the grass and crops regrow, these buildings are hidden again. For that summer, though, modern life is haunted by the people who lived, loved, farmed and fought on this land long before we called it home. It is hard not to see a message of mortality in the pale grass strips. The earth stores our past. And sometimes it can show us, if we are prepared to see it, that what we are told by those with the loudest voices is not always the whole story.

Over twenty years earlier, when Ollie was still a boy, his father, Philip, also discovered a secret in the soil. Philip had stopped being a farmer by then. When he discovered the secret, he was trying to carve out both a new life for himself and a new identity too, for when he sold the farm he lost not just the dairy and the land but who he was. Philip's father was a farmer, and his father's father too. Everyone always said farming was in your blood. But if you were a farmer with no farm, what ran through your veins then?

Philip's father had bought the farm the year after the Second World War ended, when farmers were heroes feeding a starving nation. But life moves fast and memories are short. By 1991, when Philip was a father of four trying to make the dairy pay and keep debts at bay, milk prices crashed. The smog of recession settled over Britain. The newspapers on their farmhouse table blazoned unemployment figures in panicked black ink. There were 1.8 million people without work. By the time the spring crops were sown there were 2 million. The figure would keep on rising. Philip had to feed a family on the money he got for feeding the world. But he couldn't. There was no real choice. The farm had to go.

There is a home movie of the farm auction which records the day the farm was sold: the day Philip stopped being a farmer and Ollie stopped being a farmer's son. It shows a very British day in October 1991. The sky is dull and grey. A man, white-haired, smartly dressed and wearing a maroon tie, tweed gilet, moss-green corduroy trousers tucked into wellingtons, vigorously brushes the yard with a large broom. He has the same face as Philip. One day Ollie will wear this face too. The face of the white-haired man is furrowed with lines, but his back is straight and his eyes are bright. When he sees the camera filming him he stands and smiles. The smile must have been difficult for him. He is Philip's father. This was once his farm too.

The edge of the field has been allocated for parking. Those who have come to buy the bones of the farm line up so many Land Rover Defenders that the place starts to look like a sales forecourt. Farm equipment sits in rows down the middle of the field, waiting for a new owner. Each item bears a lot number stuck onto it with tape. Old tractors. Sprayer booms. Cattle crushes. Rubber cow mats. Gates. Iron troughs. Even the outdoor privy. Everything is for sale.

As more people arrive, groups begin to form around the auction lots. They are dressed in the outfits of rural nineties Britain. Puffa jackets, double denim, feathered haircuts. Many wear glasses which make them look like different versions of Geoffrey Howe. The women wear sweatshirts in primary colours over collared shirts and stonewashed jeans or, if a little older, pleated skirts. The men wear the timeless dress of British agriculture: farmers are in blue or green overalls and landowners in flat caps, green wax jackets worn to the hip and trousers in varying shades of beige. Some of the older men have ties under their shirt collars. Everyone – man, woman, child – wears green wellingtons.

The women talk to one another in small groups. The men tip down their chins and stuff their hands deep into their pockets and nod. The atmosphere is part agricultural show, part wake. Some of the people laugh and joke and then check themselves, for although people here clearly know and like each other – and any chance for a farmer to see someone to whom they are not related should not be wasted – this is a sad business.

Philip – dark hair, kind, gentle face – talks to some of the people in the crowd. At one point his wife joins him – sweatshirt over shirt, stonewashed jeans, green wellingtons – and he takes her hand. They walk along like this for a while. The bravery of their smiles as they watch their friends and neighbours buy their life makes my throat tighten.

A bell is rung to start the sale, by hand, like a town crier. The sound makes the event feel immediately ancient, like this getting together of people on the land of their neighbour is a kind of ceremony. They are there to bid, and watch others do so, for slices of something that is more than simply property. This is a plucking apart of history. It is, in its own way, a farm funeral.

The auctioneer is dressed entirely in green — hat, jacket, trousers, boots — and carries with him a curved walking stick which he uses either to point to lots or to help him climb onto tractor bonnets, so his patter can be better heard. Afterwards, he moves into the barn. A makeshift auction mart has been built from metal gates and rows of seats from square bales. Philip stands in the home-made ring, arms out, herding a cow around its circumference. He is not smiling any more. The auctioneer, atop a platform of hay bales, calls out an uninterrupted stream of words which sounds more like a song than a sales pitch.

'Now-then-Lot-3-will-be-a-real-pearl-for-you-first-of-the-home-bred-cows-ladies-and-gentlemen-you'll-find-she'll-be-real-grand-over-the-hammer-at-seven-thousand-at-three-hundred-kilos-she-eats-in-her-lifetime-instead-of-six-thousand-and-six-at-four-one-oh-three-seven-thousand-and-two-at-four-percent-six-thousand-and-seven-last-time-currently-now-before-you-giving-twenty-four-kilos-a-day-ladies-and-gentlemen-there-we-are-now-wait-for-me-put-me-in-who's-going-to-hand-me-six-hundred-five-and-a-half-five-fifty-five-twenty-five-hundred-away-five-hundred-five-hundred . . .'

As the auctioneer continues and no hands rise from the auditorium of bales, I cannot stop watching Philip's face on the screen. I find myself thinking, *Bid, you bastards, you tight bastards; have a heart.* I should not have been so quick to judge. By the end of the decade, some of them must have had to sell their farms too.

The film ends not on Philip, or on Ollie – a small boy in overalls and boots, climbing five-bar gates and slipping on cow muck – but on his grandfather: the white-haired man. He winds orange baler twine neatly around two fingers with working hands and puts it in the pocket of his tweed jacket. A forelock of his white hair blows in the wind. His face is kind and also very, very sad. He tries to laugh back at the person on the other side of the camera but finds he cannot. As the camera focuses on his face a kind of shadow settles over his features. He is no longer smiling.

In selling the family farm when others might not have, Philip did a brave thing. It is not an easy thing to sell a farm bought by a long-dead relative when times were good and land was cheap. Often these ancestors rule over the decisions of their bloodline from the grave, binding future generations to their fate. Some sons – and some daughters, although fewer, for this is still an industry in which agnatic primogeniture can bind people's fates – inherit the farm when they do not want to become farmers or when a feeling for crops or livestock just has not been passed on along with the acreage. They do so knowing that the land is never truly theirs. Often they cannot sell it because *farmers don't sell land, they buy it*. Sometimes they cannot alter it either, for there are still old heads around the table who believe that experience trumps ideas. Many feel their farm is like another member of their family, the land being in possession of its own special flaws and quirks. Sometimes the farm is generous and loving. Each corner of it carries a memory so that the farmer feels not only as though they know the land, but also that the land knows them. For others, the farm is tyrannical and malevolent. At times they have hated it for the trap it represents, like they hate the unpunished cruelty of a playground bully. But even if there might be other landless farmers who would do a better job, each generation knows from childhood that this farm will one day become temporarily theirs. It is a farm that cost money, sweat, some blood and some tears too – even if they went unshed – and must not, therefore, be lost. It is a special kind of inheritance, one that, for some, can look like privilege but feel like a millstone.

It turns out that, in selling the farm, Philip doesn't just save his sanity. He saves his family's future too, although it takes a while before anyone realizes just how close they came to ruin. As the country sinks deeper into depression, cases of bovine spongiform encephalopathy in cattle begin to climb. BSE; Mad Cow Disease: the sickness moves through the country like an earthquake. In 1992, the year after the farm was sold, 100,000 cases are identified. It is traced to cows who have been fed meat-and-bone-meal containing the remains of cattle or infected sheep. Herbivores turned into carnivores turned into cannibals by a system that thinks it knows better than nature.

A neurodegenerative disease, it causes cows to tremble, lose muscle control or stagger. They become aggressive, nervous and frenzied. Eventually, they fall into a coma and die. Three cows in every thousand are affected. Around six million cattle are slaughtered. One hundred and seventy-seven people die from eating infected beef. It costs the UK government an estimated £3.7 billion in slaughter, compensation and related expenses.[1] British beef is banned from export to most countries around the world. Japan and America maintains the ban for over twenty years.

Then, less than ten years later, foot-and-mouth disease decimates an industry only just staggering back to its feet. When a farmer can't – or won't – shoot his condemned herd, army soldiers turn into slaughtermen in hazmat suits and do it instead. Bloated piles of dead cattle bounce along country roads in the back of overstacked trailers. People stuff towels around gaps in door and window frames to keep out the rotten stink of cattle pyres. To keep out the sound of the shots, too. Village schools close to stop children breathing in the smog. Afterwards, local people say that it takes a long time for the birds to start singing again. It is a living nightmare that will haunt farmers' dreams for decades.

Philip gets out just in time.

Philip's new job working for a wildlife trust includes the novelty of a lunch hour. He eats his sandwiches in the tangled garden behind the large house with his new office and his new job inside it. One day, sitting there, he notices something amongst the undergrowth. It is the remnants of a long-ago buried garden, lost within the wilderness. The eighteenth-century parkland had been felled for its timber in the 1960s. A commercial forest had replaced it for a few decades but now that too has been forgotten. The edges of the silted-up lakes and the broken leat that once fed a now derelict waterfall are just visible. The remains of pillars from a lost temple are smothered in greenery. Later Philip will say that, in that moment, he finally understood the hand that fate had dealt him. He knew, somehow, that finding that buried history and uncovering the ground's secrets was his real purpose.

Philip goes about restoring the garden as though it were a quest. Without telling his wife, he remortgages their home to fund the dig. And, because of him and the work he does, the garden and its buildings are uncovered, celebrated, award-winning and opened to the public for the first time in over a hundred years. Philip may not be a farmer any more, but he still knows that the soil keeps secrets.

Sometimes volunteers come to help rebuild walls on the site. They are almost always men in their forties and fifties who have found themselves broken by work and life. They have discovered that, somehow, working on the earth can prove better therapy than any gained by lying on a couch. It is as though touching the past can remind us of our connection with everything that has come before and everything that will come after us. Restoring the past does not just give us a purpose today but can make us tiny links in a very long chain. It is a hopeful thought. It is a thought some farmers have when they plant hedges or trees that will not reach maturity in their lifetime. It is the thought I have when I find myself standing at the top of a meadow field, watching an owl hunt, wondering how to fit my new present with my past, and with my future too.

Meanwhile, Ollie, who hasn't been a farmer's son since he was eight is now an engineer graduate. He sits in a swivel chair in an office with blue felt pinboards and Health and Safety posters on the wall. He wears a suit. But underneath, a farmer is buried.

When he sold the farm, Philip kept back just one field. The field has, since the farm was lost, been rented out to a neighbour. Ollie tries to persuade Philip to let him take it on, alongside his job, to see what he might make of it. According to the Rural Payments Agency, who pay farmers public money on behalf of the government, the size of the field – twenty-two acres – is technically big enough to turn Ollie into the farmer he wants to be.[2] When it is put alongside the average UK farm of 213 acres, twenty-two seems small.[3] When put alongside the larger farm holdings, which are over a thousand acres and make up 54 per cent of the country's farmland, it can seem more like a joke.[4] But Ollie doesn't think of twenty-two acres as a joke. He thinks of it as a beginning.

Father and son talk about it in the way some British men talk, where something small actually means something really big. Like when their brother dies and their eyes stay dry, but then the old dog dies and they cry and howl and rage at the sky and rock its body in their arms, and everyone knows it is really their brother they are cradling. But Ollie doesn't need his father to say the words out loud. He knows what his dad wants to say to him. *Don't do this. Do anything but this. It is not like it was. You will be a pinball in a political arcade, played around by those who will never live this life and with no chance of ever controlling the game. It is a trap. It will consume you with its unpredictability and instability. It will harden you. It will ask everything of you. It will break you.*

But Ollie cannot help it. He has the same feeling his father had when he saw the buried garden. He knows somehow that this life is his blueprint, buried underneath him. He does not want to be in a suit in a swivel chair. He wants to be outside with space around him. He wants to look up and see the sky. He wants to be in charge of his own day and of his own life. He wants to feel the kind of connection his father had felt when he watched, stone by stone, an old world being restored and knowing he was now linked to it forever. He wants to feel like he has built something. Eventually, his father concedes. In the face of passion he well understands, how could he not?

Still, he doesn't make it easy. Ollie has to understand how hard this will be. Philip will let his son rent just nine of the twenty-two acres, at full market rate and under no illusion that his father would prefer he enter any industry other than farming.

Ollie keeps his full-time suit-job but the income is not enough to enable him to farm like others farm. He is twenty-five. He has no barn or machines or fertilizer or seed or chemicals. He cannot afford them. But he does have a field. Is this enough, he wonders, to make him a farmer?

He begins by fattening Christmas geese on the nine acres, raising them on the pasture, checking on them before and after work. Then a chance conversation with the man who rents the rest of the twenty-two-acre field means he is able to take on the whole lot. He doesn't tell his father, knowing he would not agree. It is some time before the two men sit down opposite one another at a pub table to talk about the future and Ollie has to confess the deception. By then, though, it is too late. By then, he is farming.

Ollie goes to the auction mart to buy his first five cows and – delighted, excited – his grandfather, the man with the kind face and the white hair, comes with him. He insists on buying another two cows for his grandson out of sheer joy at this re-entry to a world he had thought was lost. Ollie chooses hardy cattle, beasts that will look after themselves without too much maintenance or expensive veterinary care. When you only have seven cows, you cannot afford to lose one. So he chooses native cross breeds – 'hybrid vigour' they tell him – which means that the cows have good feet, even temperaments and put on weight slowly so that the flavour of their meat is deep and delicious.

He takes the cows to slaughter in a local abattoir. His meat, fattened on his old, species-rich pasture, contains more nutrients, minerals and fatty acids than meat from cows fed grain in a barn. It is more nutritious even than the meat from cows that have been taken indoors for the last six months of their life to be fattened up on grain after a life on grass. He has read the research papers which tell him so.[5] But, he discovers, this makes no difference to the price he gets for his meat. It is – to those who buy it – simply beef.

It's hard, sometimes, to think not only that the food system is broken but also that it is too big to fix. Ollie knows he cannot fix the food system, but he thinks he might be able to fix *his* food system. He has twenty-two acres and a laptop. He can use the acres to make the food. And, with the laptop, he can start to sell it.

Ollie's local council disbanded their agricultural department long ago. Most of their council farms have already been auctioned off with vacant possession, the land bought up and the farm buildings developed into barn conversions with a library mezzanine and a double-height kitchen. You could blame austerity, you could blame cuts, you could blame the chasm in understanding between the city and the countryside and the former's view on what the latter should be used for. Whatever the reason, providing young landless farmers with a way to grow food is no longer a priority in the way it once was. Raising money is instead. And there's no better way of raising money than selling land. After all, as the expression goes, they aren't making any more of it.

| | |
|---|---|
| Application for Greenway Farm: | 125 acres plus outbuildings and farmhouse |
| Name: | Oliver White |
| Age: | 30 years old |
| Qualifications: | Agricultural Engineering Degree at Harper Adams University; Nuffield scholar |
| Farming experience: | I have spent the last five years building up a small farming enterprise. I began by rearing geese on ten acres of pasture for the Christmas market. This has grown to twenty-two acres and includes fattening a small herd of cross-breed beef cows, Poll Dorset lambs and geese for Christmas over the winter. I have also introduced table chicken in the summer. All the animals are pasture fed. The poultry is supplemented with grain but lives on the pasture in rotating coops.[6] In 2010 I launched a website selling the meat directly to customers to increase my bottom line. This has been successful. I am now confident I have the customer base to expand the business to a bigger farm. |

All of this is the truth. But it is not the whole truth. It does not explain the how, nor the why.

It does not say that, a few years earlier, when Ollie was in his job with the suit and the swivel chair, he read a book whose title told him *You Can Farm*.[7] The book said that while getting into conventional farming was near impossible without the kind of money needed for kit and chemicals or taking on a lifetime's debt to pay for them, there was another way: a low-input farm. It involved a lot more thinking but far less cash. With sunlight, water and manure, the grass would grow for free. The animals turned it into meat. The meat became food to sell. It was a cycle that had fed people for hundreds of thousands of years.

Ollie's farm tenancy interview takes place in a conference room with blue polyester carpets at the red-brick council building. Ollie watches as The Man From The Council has to persuade two others to join him, pulling them away from *more important matters*.

The farm is small: only 125 acres. Its tenancy with the council is for ten years. This is unusual. Many tenancies are now granted for only five or three years. How, Ollie thinks, with the weather so fickle and the farm cycle so slow and the soil so in need of rebuilding, can you possibly plan a life and business knowing you may have to leave it in three years' time?

When Ollie explains what he wants to do, The Man From The Council is enthusiastic. Ollie is relieved. This is his second interview for a council farm. He knows that others go through many more. Most never succeed. But The Man From The Council is interested in Ollie's plans for a farm that focuses on nature as much as it does the production of food. Could it be that he is old enough to remember living near a small mixed farm like this, which were once dotted all over the country, and misses seeing them? Perhaps it is sentiment that makes him give Ollie the lease? Or does he believe Ollie's claim that he really can turn this small farm around so that, in six years' time, he will have a full-time employee, a £250,000 turnover and a national customer base so big that he will sometimes worry he won't fulfil all the orders. I hope it's the latter. Because then The Man From The Council will be proved right.

Ollie is chosen over sixty-five other candidates. Sixty-five other people who also want the chance to become a farmer. Sixty-five other people with the desire and the passion and nowhere to put it. That's the funny thing about farming. It costs a fortune to get into and you often barely make enough money from farming alone to live off, but more and more people feel its pull. Agriculture is one of the fastest growing subjects.[8] Agricultural universities are full. There is something drawing people in: something both magical and real about working on the land that is missing from our hyper-fast, hyper-connected, hyper-lonely, hyper-empty lives. Maybe the thing they want to find is purpose. Maybe they think the land will give it to them. Maybe they are right.

Ollie knows there are sixty-five other applicants for the farm because, when he is chosen as the new tenant, the council remind him at every available opportunity. They remind him of it when he arrives on the first day of his tenancy to find the farm derelict. It is a shattered skeleton of a place, a daily reminder of what farming looks like when it all goes wrong. The previous farmer had tried to run it as a dairy but had left in the middle of the night with his rent unpaid and a year left on the lease. The knackerman tells Ollie that the farmer could afford neither to feed his cows, nor pay to have the ones who had starved to death taken away. There are animal carcasses rotting under hedges and bones lying in ditches. The dairy parlour and tank room have been stripped of every kind of metal that might be sold for scrap. Any equipment that could be ripped out has been taken. Each corner turns up mounds of rubbish. Exposed electrical cables hang from the walls. The gale-breakers on the side of the barn have unfurled and been shredded by the wind. Every single shed is broken. Blackthorn and hawthorn suckers sprout up all over the field. Broken fencing has been swallowed by hungry hedges. There is no running water. Even the farmyard gate is missing.

For the last six months of the old tenancy the council have rented the farmhouse out to someone else. Ollie discovers that the temporary tenant worked as a vermin exterminator, who fed the rabbits he shot to the dozen dogs with whom he lived. This explains why the freezer, which has been switched off, is filled with tray after tray of rotting meat. There is no central heating. Dog shit on the floor. Holes in the wall. Damp wallpaper hanging down in peeling strips. And, in the middle of the conservatory floor like some primeval omen, a deer's liver.

Still. Sixty-five other applicants. He is lucky to have it at all.

It is hard at the beginning: very hard. Nothing works. But sometimes, when nothing works, it is easier to begin a transformation. Like in life, it can be necessary to reach rock bottom before the effort of change becomes worthwhile.

Ollie begins to overhaul the farm. He begins with the soil. He has read and seen enough to realize that the key to it all lies beneath his feet: the earth. What he grows in it, how well it feeds his animals, how fertile and healthy he can make it, all this will determine whether he can get this farm to work. The money he has saved from his office job is enough to get him started. But where other farmers might have invested in new equipment, the latest time-saving technology or newest, flashiest tractor, Ollie buys something else. He buys seeds.

He doesn't buy just one kind, either. Instead of the green ryegrass that other farmers sow, Ollie drills bird's-foot trefoil, sainfoin, chicory, ribgrass, burnet, yarrow, sheep's parsley, six different types of clover and four different types of meadow-grass. Had some of these grown by themselves, their seeds blown in on the wind or trampled by hoof or boot, other farmers might have called them weeds and sprayed or hoed them away. He sees it differently. These plants are forage.[9] Each one contains different kinds of minerals and nutrients, which Ollie hopes will both keep the vet away and transform land depleted by years of maize grown to feed the former tenant's dairy cows. The plants' roots will form a network of connections with one another beneath the ground, sharing requests and offers of minerals, nutrients, carbon and water via the fungi which join them together in a kind of deep-earth telephone network. *Give me this and I'll give you that in return.* It's a huge underground commune, working in perfect harmony right beneath our feet.

Ollie knows he has to make this ground work hard. Unlike the latest self-driving tractor or 50,000-flock chicken barn leveraged with borrowed money, these plants do not depreciate in value every day. They do the opposite. These plants are his insurance policy. If the plan works, they will feed his animals through flood and drought while other farmers' fields clog and crack.

He builds electric-fence routes around the farm to make it easier to move the animals onto fresh pastures more often. He will graze the cows and sheep together, making the animals easier to move. The sheep will eat the plants that the cows don't like; the cows will eat the plants that the sheep don't like. If he thinks he might run out of grass he can slow down the rotations, encouraging the animals to eat every type of plant rather than picking off their favourites. By moving the cows and sheep regularly, preventing them from going back onto grass they have already grazed, he will stop them eating the new fresh grass shoots before the plant has had time to establish itself. This way of grazing means he can double the amount of forage in his fields.

It also means the animals will not go back over areas

where their old muck lies. The parasites within it will die without being able to reinfect them, breaking the cycle, so he doesn't need to pay to worm them. And if he doesn't worm them, dung beetles and parasitic wasps – killed off by artificial wormers and fly treatments – will return to live in and eat the dung. The parasitic wasps, barely big enough for the human eye to see, lay their eggs inside fly eggs and, when the wasps hatch, eat the fly larvae, so fewer flies bother the cows and sheep in summer. Dung beetles break down the muck in days rather than weeks, incorporating the natural fertilizer into the soil, attracting worms whose tunnelling aerates the earth, and further stopping flies from laying their eggs in it.[10] When the dung beetles return, so do the creatures that eat them, and so do the creatures that eat the creatures that eat the beetles. Thus the cycle goes round in a hundred small ways, each link in the chain affecting something entirely unexpected.

Finally, Ollie connects the water, linking the drinking troughs in the fields with an overland pipe that can be taken to portable water troughs and moved to wherever the animals need it. The sheds are repaired. The farm-yard gate is replaced. The hedges cut and the broken fences within them mended. He sets about mending all the evidence of what happens when a farm fails while trying to push aside the thought that, one day, the same might happen to him too.

Ollie scowls at the cracks in the ground, cut into the earth by the May drought. He is wearing trainers. There's no need for boots. There hasn't been mud on the farm for weeks.

In 2018, when skeletons of buried buildings were burned onto the ground, people treated the drought as a one-off. Newspapers showed pictures of red-skinned people packed onto beaches, lying alongside each other like rows of snapper for sale at a fish market. The farming press showed parched crops withering; wheat with its flag leaves curling to protect the crop; grasses yellowing; chunks of pasture with nothing left but cracked brown earth. Both shouted about BRITAIN'S HOTTEST DAY IN HISTORY!

It turns out it wasn't a one-off. It turns out that's the weather now. There isn't really any need for capitalization any more.

Now it is 2020. Only three months earlier the papers were yelling UK SUFFERS WETTEST FEBRUARY ON RECORD! Now they shout UK EXPECTS HOTTEST DAY EVER![11]

This time, though, the photographs are different. The world is emptied; its citizens shut up. Pictures show not tourists sunbathing but places emptied of the busyness which defines them. Times Square; Mecca; the Las Vegas strip; the Champs-Élysées; Piccadilly Circus. They are as empty as a set from a disaster movie.

It is only the farmers who are watching the weather. Everyone else is watching the progress of a contagious plague sweeping the world, touching every continent but Antarctica. For most people, life as they know it has stopped. New words are entering everyone's vocabulary.

Now people are locked down, furloughed, self-isolating. Now Ollie is a 'key worker'. Now there is no such thing as normal for anyone except farmers, because no matter what is happening in the world the crops and animals keep on growing and people keep on needing food. Farmers' lives have – more or less – just carried on. For them, legal isolation doesn't really feel that different from the usual kind, though few had predicted that now the whole country – and most of the world – will finally understand the kind of toll this can take.

The last few months have brought other revelations. As soon as the lockdown rules relaxed, farm footpaths were filled with more walkers than anyone had ever seen before. Farmers spent their evenings disinfecting gate-posts and stiles in case the rumours were true and Covid-19 could linger on the wood. They hung up signs warning 'GROUND NESTING BIRDS: DO NOT TAKE DOGS OFF LEAD' and returned to find them hidden under a collection of plastic green dog-poo bags, hung by their handles on the post like baubles for someone else – who, though? – to collect. There was litter: everywhere.

The farming press was filled with reports of dog attacks on sheep and new-born lambs by animals whose owners thought their pets civilized, quasi-human, one of the family. Every farmer knew of someone whose ewe had aborted after it was chased or had a sheep mauled so badly it had to be shot. One farmer lost several sheep when his flock trampled one another to death after being driven into the corner of a fenced field by a barking dog. Many farmers were glad their children were off school to help collect escaped animals after field gates were left open. Tensions between those who saw the countryside as escape and those who saw it as their home escalated. Dawn raids were carried out by police trying to catch city people renting cottages after villagers had shopped them. At the edge of the Brecon Beacons in Wales a white van was parked diagonally across a country lane with 'FUCK OFF BACK TO LONDON' written on it in black paint.

But it seemed like city people were suddenly desperate for the countryside: gasping for green and birdsong after weeks of walking on city pavements and around the same tired park. There was a surge in books and programmes and podcasts about nature. The daily reports of death figures were set against news of a revival in gardening, vegetable planting and interest in the natural world. People kept talking about an awakening, as though it took millions of people to die before the rest realized they were alive. Maybe, sometimes, that is what it takes. Maybe death can enable life after all.

Ollie kicks at the ground with the toe of his shoe. By the end of the month, UK soils will be declared the driest on record. He worries that if there is no rain then the sward won't grow. He worries he will have to use precious winter hay to feed his animals, as he knows other farmers have already done.

He is standing in the field at the top of the valley that rises away from his farmhouse. In the field below this one are rows of long, rectangular cages Ollie has built. Each contains just over seventy chickens sheltering from the sun in the shade cast by the coop in the cages' centre. The cages are on wheels so that, each day, they can be dragged onto fresh pasture for the birds to peck and scratch for worms and insects. When the cages reach the top of the field it is easy to look down and see the bright green oblongs where their muck has fertilized the ground surrounded by the pale green strip where their cage has rested. To his right, in the field alongside the farmhouse, a flock of white geese waddle and honk. In the field in between stands a 'gobbledygo' – a triangular frame on wheels that Ollie has built, and where, in August, young turkeys will roost.

He began the free-range poultry enterprises soon after he arrived. He had a few years of peace before the foxes learned of the buffet living near them. Then it was carnage. The foxes would dig a gap under the wire big enough to take a chicken's leg in their mouth, rip it off and leave the bird inside to die. Ollie woke many mornings to find up to ten legless birds dead alongside the terrified living ones. If he managed to shoot the fox there would be another to replace it before the month was up. The badgers were worse, though. In one orgiastic slaughter a badger – protected from Ollie shooting it by law – dug its way into the pen and massacred sixty-five chickens in just one night.[12]

These kinds of predators aren't a new problem for those who keep chickens. Farmers, country dwellers, even city-living human-rights barristers dressed in their wife's kimono have found them so. The solutions are usually limited. Keep the chickens inside, build Fort Knox, or keep killing the predators when you can. But the way Ollie is farming is different. He does not see himself as standing on top of a pyramid, trying to control everything underneath him. He sees himself as part of a circle. He is a predator within the circle but he knows that his decisions have an impact not just on the focus of these decisions but also on everything else attached to them, and so he learns both the part they play and their connection to the rest of nature too.

'*Neigh-chure*' the other farmers mock. 'You wanna tame it, mate, not give it a cuddle.'

But Ollie – check-shirted, jean-and-boot-and-fleece-wearing, thirty-year-old agricultural graduate from three generations of farmers Ollie – doesn't want to give nature a cuddle. Nature is not cuddly. Nature bites. She invades. She colonizes. She can poison, rip flesh, burn skin. She can kill. And not just in her howling lashing thunderstorm form either, as any farmer with an acorn-eating cow will tell you. This kind of farming is not about cuddling nature. It's about respecting it. It's about understanding it. It's about living with it in your eye: seeing yourself inside the circle rather than outside it.

So Ollie gets some dogs. They aren't the kind of farm dogs you usually see. They are Maremmas: an ancient breed that dates back to the Roman era. They were once used to protect livestock from wolves. They are huge and white and fluffy, like a golden retriever on steroids. Their low prey drive means they do not want, as most dogs do, to hunt and kill. Instead, fixed within their DNA is a drive to protect whatever they decide their charge is. In this case, their charges are the poultry. It is a strange thing to see a pair of dogs lying in the sun, ignoring the small flapping birds which surround them, but it is not exactly new – guardian livestock dogs are mentioned in Virgil's *Georgics* – or, increasingly, that unusual. They are used widely in Australia and parts of Africa. In Italy and Switzerland, hill shepherds routinely use them to protect their sheep from wild wolf and lynx and, in Georgia, America, one farmer uses these dogs to protect his free-range animals spread over 2,000 acres.[13]

When Ollie collects the two puppies he learns that every other buyer is a poultry farmer. The breeders, farmers themselves, says they have discovered another accidental consequence of keeping the dogs. In Cornwall, a county plagued by the grim spectre of TB, often spread from farm to farm by badgers, they are one of the only farms in the area that has been clear for years.

It is not all perfect. It never is. Before the dogs came two alpacas, destined to do the same job. Instead of guarding the poultry they ran – long-legged, long-necked, ridiculous – in the opposite direction when anything – fox, badger, Ollie – came near them. More problematically, the farm is criss-crossed with public footpaths, and Ollie finds himself at the wrong end of complaints from local walkers when the dogs are jobless between poultry seasons, and the farm itself becomes the focus of their guardianship.

Regenerative farming – for that is what Ollie's way of farming is called – is not problem free. There is always a compromise or a lesson waiting. It is a labour-heavy system in an industry that is mostly mechanized. Ollie must rely on volunteers who want to learn this kind of farming during busy periods.[14] Even so, some say regenerative farming can do more than produce food. They say it can restore insect populations that are so in decline that studies have warned nearly half could die out completely in the coming decades, and in turn restore the population of birds and other creatures that feed on them whose numbers have crashed without the food and habitat they rely on. They say that the way animals are grazed in this system does not just enable plants to survive droughts and floods but ensures carbon is locked up as organic matter, cooling the planet in the process. That it increases soil biology damaged from overuse of artificial nitrogen, pesticides and deep ploughing, which in turn has led to erosion, soil loss and water and air pollution. That keeping living roots in the ground all year long helps to keep rain within the fields rather than allowing it to wash off and flood the villages downstream. They say that in winter, where other farmers' land is boggy and poached and the roads are streaming brown with soil run-off, you can walk the length of a regenerative field without picking up mud on your boots and watch as the drains run clear and clean. None of this is mythical. I have seen it, both on Ollie's farm and others, for myself.

Some say this way of farming is about more than

growing food.[15] That it is a movement which can repair not just ecological ills but social ones too. That it asks us to look at our connections with the planet and other people differently. And that, unlike almost every other industry, farming has the power and potential to restore the earth rather than merely stop damaging it. Some say that this way of farming has the power to heal.

As Ollie kicks the earth with the toe of his trainer, he cannot yet say whether his farm is healing. All he knows for sure is that, even though the sun is beating down on everyone, he still has grass where others do not.

That afternoon he checks the freezers in the small shop he has built in the farmyard by the roadside with a self-pay system for walk-in customers and click-and-collect for those who have paid online. Nearly every single freezer is empty. His customer numbers have swelled beyond anything he could have imagined as people realize not just that they cannot always rely on the supermarkets for their food, but that maybe they don't want to any more.[16]

In 2025, in only a few years' time, Ollie's tenancy will be up. His soil tests have already confirmed that the way he has farmed has significantly improved the council's land. His hedges, cut on a rotation every two years so that there is always winter forage for birds, are thick with wildlife. Any money Ollie has made has gone back into the business: building it up, diversifying it into as many areas as he can so that if one part of the farm has an unsuccessful season, he will still manage to make a profit. By the time his tenancy is up, Ollie will have lost over half of his public subsidy as the phased government withdrawal of the Basic Payment Scheme begins to bite. But the business he has created here is resilient. He has sown the land with resilient plants that survive drought and flood. He grazes resilient animals that rarely need his attention and assistance. He sells his food directly to customers who want to know what they are eating, and what they are eating ate, too.

After one of my visits to his farm, Ollie tells me he has learned it was bought by the council after the First World War. His great-grandfather had also taken up the offer of a council farm when he returned from war, not far from where he is now. All over the country these farms were offered to veterans to give them an occupation, and so a purpose and also, very possibly, a place where the mental and physical scars of war might heal. When he tells me this I think of Philip and his garden, of Ollie and his farm, and of those who tell me that farming this way does not just bring resilience against weather or spikes in prices or cuts in subsidies, but that it can bring healing too.

I don't know whether the council will really understand what Ollie has done here or how much he has improved their land. He was the last farming tenant they took on. Their current policy is to sell their farmland. When other farm tenancies have come to an end, this is what they have done. Ollie may never own this land. But he also knows that it is this land that has made him a farmer.

Some might call his farm niche. *Niche*. It's supposed

to be pejorative. There is an implication of smallness, of something uncommercial or unworldly. A bit of a joke compared to Proper Farming and Feeding the World. But the word niche means to find a place, employment, status or activity for which a person or thing is best fitted. It comes from the Middle French *nicher*, meaning 'to make a nest', and is derivative of the Latin for 'nest'. To be niche is to slot into a place that fits you best. To find your nest. Ollie has built the farm that fits him best and where he is as protected as he can be from the threats that other farms face. I cannot think of anything less pejorative than that.

*My bugbear is, why are tidy people right?*
*I mean, what in the world made them always right?*
*Because it's all bollocks . . . it's the worst thing you can do!*
*The best thing you can do is bugger all and just leave it.*

Paul, farmer

# 5. Year Two – Looking Underneath

When the new year comes, we start with the pasture fields. We spend Friday afternoons walking their edges with the children on our backs, swaddled in hats and gloves. We need to plan the future of these fields now the responsibility of their future lies with us. We have the privilege of knowing this will not be our sole income, but it must do more than wash its face. We want to see if we can make this farm both ecological and economical.

In February it snows again. We light the wood burner to warm the darkness of the afternoon and spread out paperwork on the kitchen table while the children nap upstairs. There are copies of old maps and I lean over them, trying to orientate myself. One, from the Records Office, has the old field names on it. Another, an 1884 Ordnance Survey map, shows not just every field but also every pathway, waterway, building, woodland and hedge. It takes me a moment to understand that the small wavy circles drawn onto the map are not a general impression of trees, like I've seen on modern architectural drawings. They are pictorial representations of actual, singular trees. Each one is plotted, small or large, and edged with a bushy line. I stare at them and then out of the window at the meadow field and wonder at how these two places – the one out there and the one on the table – can be the same. Those three pasture fields that sweep away from the cottage and down into the valley, the river running along their far edge, were once six. The hedges that once divided them were so old they were likely to date back to the Bronze Age. Ben uses other online maps to work out that somewhere between 1956 and 1977 over a kilometre of hedgerow was removed from this small patch of land. More was lost in the mid-1980s. With the hedgerows went at least forty hedgerow trees as well as a small copse along the riverbank, which was then ploughed up.

Unlike most parts of the world, two-thirds of England has had a continuously hedged landscape. We have had hedges for six hundred years or more, long before the enclosures.[1] Some hedges in England date back to prehistoric times and most were well established by 1400. The word 'hedge' is even older. It appears in Old English, German (*hecke*) and Dutch (*haag*) and translates as enclosure (as in the name of the Dutch city Den Haag, The Hague). Our old hedges had to be maintained on a rotation each autumn to ensure there were no gaps big enough for an animal to escape through. Their bushiness meant they became the homes of voles, moles, dormice, hedgehogs, bats, snakes and many species of birds and other creatures. They provided the animals who grazed in the fields additional forage and shelter from wind, sun, rain and snow. But when the country was dug up to feed people after the Second World War, hedgerows were taking up space which could otherwise be used to grow food. New machinery made cutting or digging out hedges much easier than it had ever been. Flailing (machine trimming) rather than laying (cutting through the stem of each plant, bending it over and interweaving it between wooden stakes) became standard practice. Hedges were pulled out at a rate of 9,500 kilometres a year.[2] In many parts of the country, well over half were lost.

Farmers were advised 'If in doubt, grub it out' and continued to receive payments for taking out hedges right up until the 1970s, long after the country stopped starving. Bigger fields meant easier production using equipment that couldn't have navigated around the old hedgerows anyway. By then there was fencing. Wire and wood would do the job more cheaply and effectively than a slow-growing, time-intensive hedgerow ever could. When I am told the stories by those who remember it happening they often speak about it with faint shock or surprise, like they are remembering an accident in which they hadn't understood that someone died. One farmer remembers being assured that it was cruel to leave hedgerows by the road for birds to nest in for the fledglings would only be killed by passing cars. Taking them out was doing the birds a kindness.

Many of the hedgerows left were not maintained because few had the time when the focus was on producing food rather than the fuel the wood from these hedgerows once provided. There were not many farmers who could afford to keep on skilled labourers to do the hedging work as their fathers had done. As generations passed, these skills were forgotten. Maintaining hedges in the autumn is a hard physical job. Filling in gaps or splitting trees to bend the branches and weave them into a living wall cannot be mechanically outsourced. It is back-breaking work, with forearms and faces left scratched by thorns and branches. It's not surprising, then, that with alternatives available that don't leave you scarred and bleeding, almost half the loss of hedgerows in England was caused by lack of management.

There were other reasons too. Big hedges were thought to shade a crop from sunlight or compete with it for nutrients and water and it would be years before research found that hedges formed a relationship with the plants inside the field. It was not just that they would protect the grass or crop from being battered by wind or rain, or that their roots helped to stop water or soil leaving the field. They also formed a biological connection underground. The hedges' deeper roots brought up water and nutrients, making them available to the shallower-rooting grass or crops in a strange phenomenon so clear that it can be seen on satellite images and has been nicknamed 'hedge edge'.[3]

The law, as it can be, was slow to catch up. In 1981 it became illegal to damage the nest of any wild bird while that nest was in use or being built. New rules meant farmers could no longer cut or trim a hedge between 1 March and 31 August for the same reason. In 1997 regulations made it a criminal offence for anyone in England and Wales to remove or destroy certain – although not all – hedgerows without permission. But, as with all law, there are shades of grey within the black and white. An old established Suffolk hedgerow half an hour from our farm was filled with native trees and shrubs. Locals in the village would pass it daily. Some loved to listen to the song of the nearly extinct turtle dove that nested within it each summer, a bird whose population has fallen by 97 per cent since the 1970s. They woke one autumn morning to find the hedgerow flailed to the ground. It was then flailed again the following February to remove any woody regrowth. When the villagers complained, the local council sent an officer to inspect. The officer found some regrowth in what was left. Those green shoots meant, they said, that no offence had been committed. The hedge had not been legally destroyed even though its integrity, its usefulness, its life and the life within it had been.

We decide to put the old hedgerows back. Each one will slice a triangular chunk off the corner of a field. Because we are new at this and unconfident, our local wildlife trust tells us what we should plant. Our hedges will be filled with percentage specifics, chosen for the exact region: 50 per cent hawthorn, 25 per cent blackthorn, 15 per cent field maple, 2 per cent holly, 2 per cent wild privet, 2 per cent guelder rose, 2 per cent dog rose, 2 per cent buckthorn. They must be protected from the dozens of rabbits who burrow along the edge of the lower field by expensive wire stapled to wooden posts, or the rabbits will have the lot eaten by spring. The majority of this cost is to be covered by a grant from the government, who, some fifty years earlier, paid a grant to have them taken out.

As Ben writes down the names of the trees we want to plant in our application for Countryside Stewardship, I look up their pictures online. Field maple. Hornbeam. English crab-apple. Common oak. Alder. Downy birch. Black poplar. This last native tree was once a common one. It is now Britain's rarest, with only 600 wild female trees left. It is a tree so ancient it appears in Greek mythology: Phaethon's sisters, mourning their dead brother after he was killed by Zeus, grieved so deeply and for so long that they became black poplars shedding amber tears on the banks of a river. The tree grew so well in boggy conditions, living up to 200 years, that it was once planted to mark field boundaries on floodplain land. Its wood was flexible and robust, used for cart wheels, wagon bottoms, clogs, poles and fruit baskets. It was bendy enough for Victorians to use its young shoots as clothes pegs. It has a deep fissured bark, thick and knobbled, with huge sweeping branches and leaves shaped like a heart. I read that John Constable painted it in *The Hay Wain* and look up the picture on the National Gallery website. I see it differently this time, not as a cliché from the lid of a biscuit tin but instead as a record of something that once grew just thirty miles from where I am now, and which has been almost entirely lost.

Replanting the hedgerows allows us to reclaim the old field names, which we take from the old maps. Each name tells us something about the field's former life. It feels like we are unpicking a past, looking for clues as to how this land should be used in the future. Here is the field where sheep were once dipped, for it runs next to the river and the tradition of doing so was so ancient that the field name, 'Sheepwash', has its root in Old English: *sceapwoesce*. Here was a field called a 'Brick Works'; there the nineteenth-century kiln that lent its name to the small strip of field that lies on the other side of the river alongside the road. Here, in the middle, was an 'old sand pit' which has left a dip that spring rain collects in and where the snow lies deep enough to reach the childrens' chests when they jump into it. Here is the field that takes its name from the 'Broad Oaks', three of which still flank the horizon. Here is the field whose old name, 'Great Fen', meant it was once pre-drained marshland, which must be why two great white willows, whose roots so love rivers and wet woodland, have grown up along its edge, their leaves so small and shimmering that when the wind blows through them they look like their branches are covered in pale green sequins.

Sometime later the son of a local farmer puts us in touch with a tree-surgeon friend. He is to cut down some overhanging branches from a dead ash before a strong wind knocks them onto the road below. When we are told his name we realize it is the same as a field name we have just recovered. When the hedgerow was lost, so was the field's original name and its history too. The tree surgeon tells us that his family has lived and farmed in this area for a long time. So our new/old field now shares its name with the son of the son of the son of the son who may once have worked upon its land. Now he is working upon it once more. It feels like we have found a link we didn't know was lost and joined it up again.

'Here' says Ben, pushing his laptop at me across the kitchen table. 'You can read the guidance booklet . . .'

It is spring. Spring now means we have deadlines for more forms than I've ever seen before. Ben and I have both worked in the public sector and are well used to government paperwork. This is on another level.

We are registering to convert the pasture to organic grazing.[4] We have found a neighbouring farmer, Martin, who will eventually merge his herd of Red Poll cattle with our three lonely cows and manage them all together, rather than the part-time guy who had previously looked after them. Martin is an organic farmer and the rules mean he cannot graze our land with his cows until that too has been converted to organic.

It is an easy decision: no fertilizer or cultivation has taken place on the pasture for years. We will charge Martin a small rent to put his cattle and sheep on the land but later, when I see how the grasses and wildflowers change and respond to the grazing and muck left by his animals, I wonder if he should be charging us.

We are also applying for Countryside Stewardship: a five-year government scheme which covers the cost to farmers of managing their land in a way that is beneficial for the environment. I scroll down the document on the screen. It seems to go on forever. I look at the number of pages.

'It's 123 pages long!' I look at Ben.

He grins. 'Yes. It is. And don't forget the other manual . . .'

'The *other manual*?'

I click back. *Mid Tier options, supplements and capital items.* I open the document. It is 312 pages long.

'Have you read all of this?'

'Pretty much. Test me. Name a code, any code.' He gets up and goes through to the kitchen.

I scroll down the list. There is an index of codes with descriptions like 'skylark plots', 'beetle banks', 'unharvested cereal headland' and 'autumn sown bumblebee mix'. There are nearly 150 of them.

'Um . . . AB8.'

'Easy. Wildflowers.'

'Nice! AB9.'

'Also easy. Seeds for winter bird food. Non-organic. OP2 is Organic. Although now I've done the stewardship form I have to do the Basic Payment one. Which has, of course, a completely different set of codes for each option.'

'What?'

'Yep.'

'Why?'

He stands in the doorway and looks at me, deadpan. 'Because: government.'

Ben spends hours filling out the forms. The Stewardship application is competitive: there is no guarantee we will get it. He shades up maps with a variety of hatching, fills out spreadsheets, writes up narrative guides for what we want to do with letters of support from wildlife trusts and heritage organizations. I pretend to help him.

Afterwards, when I print off all the paperwork, a memory returns. Ten years earlier, when I had not long qualified as a criminal barrister, I was waiting for my case to be called on in the back of a local magistrates court. The courtroom was in a 1970s building with an interior as breeze-block grey as its exterior. I sat on one of the fold-down seats in the courtroom gallery and watched the case before me while I waited. It was an RSPCA prosecution of a farmer. I could see the defendant sitting in the dock. He was somewhere between sixty and seventy, thin, with grey hair frazzled around his scalp. He was wearing a tweed jacket patched at the elbows, a checked shirt with a tie and brown trousers that were too big for him and bagged where he'd belted them tight at the waist. On his left hand he wore a gold wedding ring, which he fiddled with, turning it around and around. It was the only bright thing about him. I wondered what he made of us lawyers sitting on the fold-down seats in Andover Magistrates Court – also tweed, also threadbare – with our black suits and shining polished shoes and blank expressions. I could take a guess.

The farmer was represented by a barrister in her twenties, as I was then. I imagined he had got legal aid to pay her. Had the case happened now I suspect he'd have been on his own, as legal funding cuts would have meant he'd have had to defend himself.

There was to be no live evidence. The farmer had pleaded guilty to multiple charges of neglecting his sheep. He was there to be sentenced. The prosecutor read the evidence out from his statements in a booming voice, pausing for emphasis. He handed up big glossy photographs to the magistrates after describing their contents. One showed the bodies of half-decomposed sheep discovered by some walkers. Another was of a wooded part of the farm where the farmer had tried to bury dead lambs in a shallow grave. The prosecutor described the squalor of the animals' living conditions, the sheep that had to be shot because they were so lame and half starved, the ones whose living flesh had been eaten into by maggots and flies because they had not been shorn, the lamb corpses with eyes plucked out by rooks and crows. I watched the magistrates look at the photographs. They looked like most of the magistrates I have appeared before: middle class, white-faced, white-haired. They narrowed their eyes and tightened their mouths into thin lines of condemnation.

As the prosecutor finished summarizing the facts of the case I watched the farmer's shoulders slump. Then his barrister stood to give her client's mitigation. He had not been coping well, she said. His son, who usually helped him on the farm, had been in an accident on his quad bike and was now unable to work. They had no other workers: they could not afford to pay wages and the farm was isolated. The local bus now only stopped nearby once a week. The farm administration and paperwork had mostly been done by the farmer's wife. She had died suddenly of cancer three years earlier. He admitted he was not managing, that he tended to put his head in the sand when there were problems. That he loved his animals. That he took pride in his work. That he was a good farmer, on a farm that had been passed down to him over three generations. That it was in his blood.

Her declarations sounded weak next to the pictures of rotting sheep corpses and the statements from the RSPCA inspectors. They had found broken gates, dirty water troughs, carelessly stored chemicals and tools. Their visit to his farmhouse was spelled out in humiliating detail. Piles of paperwork and unopened letters. A sink full of washing-up. Overflowing bins. Empty cupboards. Three dogs chained up outside, thin and barking.

I could not take my eyes off the farmer. He looked like those I used to see at agricultural shows: one of the old boys who just wanted to go back to the way things were when people left them alone to farm the way they always had done. Back then I felt sorry for him, but I had not fully understood until now how easy it must have been for him to sink. Now I watch reams of paper spew out from our printer and wonder how he could ever have found the time, or resources, to interpret manuals hundreds of pages thick with sentences I had to read three times to understand, or print off old maps from record offices, or email people for letters of support for the application, or find a way to apply for the public money for which he may have been eligible and which might just have helped.

I remember the expression on the face of the middle magistrate – the Chair – as she looked over the farmer's head as he stood to receive his sentence. Twenty-four weeks' imprisonment, suspended for twelve months, together with a community order to complete 200 hours of unpaid work and a fine for costs of £215. Afterwards my client, a serial burglar with the gift of the gab and an intricate knowledge of the legal system, was sentenced to half that length.

Later, at the station, I saw the prosecutor standing on the platform waiting for his train back to London.

'Always get a result in Hampshire,' he said to me, laughing loudly. 'Think more of animals than they do of people, this lot.'

That summer we take on the care of a brood of chickens. We house them at the end of the garden, in a large coop that sits within some crab-apple trees, which fall to the ground in great heaps and which I hope the chickens will eat. They don't. There are five cockerels: too many for too few hens. I worry that our neighbours will begin to hate us as they compete to call in the sunrise. We decide we can only keep the father. We call him Harry, because it is the name of someone cocky. The four younger cockerels must be killed. We make the resolution and decide if it is to be an honest one we must do it ourselves. If Ben kills them, I will cook them. We are country people now.

We look it up in my new chicken book. It tells us to do it in darkness, when the bird is calmer. There is a photographic montage with instructions alongside each picture. It makes the killing look similar to building an

IKEA flat-pack. When it is dark Ben goes to the coop to catch a cockerel. He is out there long enough to make me worry about the length of the bird's leg spurs. Just as I am about to go and find him he appears in the dark, a cockerel clamped firmly under his arm. It looks gigantic. I can tell Ben is nervous. As he holds it I suddenly realize that I haven't really killed a living thing on purpose before, and certainly not like this. We go back and forth, both of us realizing what we have promised to do but knowing we cannot back out now. I hold the book up for him to see and turn my head away. There is a lot of wrestling then I hear the crack of bone. When I look back the cock is dead and Ben – who has killed rabbits, pigeons, pheasants and other game before – looks shaken.

We pluck and gut the bird. It takes hours. I slow cook it in an old stoneware pot designed for coq-au-vin but the moisture escapes from where the ceramic top and bottom don't seal properly and the meat is stringy and chewy and disgusting. We ask someone else to kill the other three cockerels. Cowards both, we are not there when he does it.

I love the hens. I love them even when winter comes and they stop laying and their coop is dark and a chore to clean out and it is full of spiders ready to tangle in my hair. Each breed has a different personality. They are curious and funny, and when I call them they waddle over, greedy for treats, like a gaggle of drunks.

One day I am filling up their feeder. It is dusk, and most of the hens are already sitting in the laying boxes.

Seeing them lined up reminds me of another kind of chicken hut: one I saw when I was twelve.

'Want to go into one of the sheds . . . ?' my friend Sophie had said. It was the long summer holidays and we were bored and messing about in her grandfather's farmyard. It was almost completely empty except for a few big sheds and some machinery in an open-sided barn. It felt eerie in the midday sun, hot on our faces. I turned to look at her. She looked back and grinned. I knew the question was a dare.

'All right.'

Sophie looked to check no one was watching, then slid back the shed's massive steel door, leaning on the handle with all her weight. To begin with I could see only darkness. Then, slowly, rows and rows of chickens came into view. They were piled into cages on top of one another as far as I could see. They seemed to reach all the way up to the roof. The noise and smell of them made my heart race.

'I dare you to run to the end,' Sophie said. 'Go on. I've done it.'

She knew I wouldn't say no. I stood alongside her at the entrance of the shed, took a breath and began to run. The hens erupted into noise, crowding towards the front of their cages. Beaks and claws seemed to be trying to grab at me. I tried to focus only on the end of the barn. When I reached it I slammed my hand against the concrete breeze block. I turned back and could see Sophie silhouetted in the open doorway. The noise and smell and heat were overwhelming. Adrenaline washed over me and for a moment I wasn't sure I could make it back. I put my head down, looking only at the doorway and started to run. When I reached the entrance I spilled out into the yard, panting and blinking, bending over with my hands on my hips. Sophie pulled the large shed door shut behind me, grinned and gave me a look: the kind you give someone after you've taken them to see a horror movie.

Battery farming – keeping hens in cages smaller than an A4 sheet of paper – was banned in 2012, nearly twenty years later. But whether kept as broilers – chickens bred for meat – or layers – chickens bred for eggs – chickens in Britain remain big business.[5] The UK poultry sector is amongst the largest in Europe. We eat over two million chickens a day. Five companies account for 90 per cent of the 19.5 million birds slaughtered in Britain each week and supply all the big retailers as well as chains and fast-food brands.[6] But their chickens are very different from mine. They are not bred to be curious, or friendly, or flighty. They are bred to grow. In 1925 it would take a chicken 120 days to reach slaughter weight. Now it can take just thirty-five.

Later I will stand in a shed filled with thousands of laying chickens pecking at my feet. They are inside because of avian flu and I am wearing a hazmat suit and a protective hat and paper socks over my boots. The farmer has built an extensive free-range egg unit after a long-running battle with the local villagers who do not want him to do so, although it's impossible for anyone to look inside their fridge and see if they are hypocrites. The shed is huge, which is lucky, as the birds must stay here until they are legally allowed outside again. Times have changed. Free range, says the farmer, is now the entry level for eggs since it became a legal requirement to label where they came from. Consumer power alone, it seems, can be enough to change the market.

Standing amongst so many thousands of chickens is claustrophobic. I think of my hens at home and want to turn away. When we leave the barn and step outside into the sunshine and air I am relieved. It is a top-of-the-range shed with all the clever lighting and perches and toys to keep the hens happy while they wait to be let out again, but it's still hard not to be overwhelmed by how unnatural it is. But while we still want to buy chickens for less than the cost of a magazine and millions of cheap eggs all year round, how else – I think – does anyone really expect this to be done.[7]

It is Wilfred who finds the bodies first.

'Mama, there's blood,' he says, and my stomach drops. But the blood is not his or his brother's. A fox has dug under the chicken wire. There are feathers everywhere. Three hens are dead. The decapitated body of one is sprawled out in the middle like a crime scene. Another lies on her side, half of her stomach eaten. Her guts spill out like pale fat worms. The fox has not just killed for food, but for fun.

The hens cluck around their dead sisters' remains as though they know what has happened. I look for Harry and cannot find him. Then he comes round the corner of the shed as though he had been hiding. I see there is blood on his white breast and on his spurs. At first I think he has been wounded then realize he is the one who has attacked. The blood belongs to the fox. The children stand in the gateway, staring at the gore. Then, after a moment, Wilfred wanders over to poke at the hen's intestines with a stick and ask me what fits where.

Later that summer we move the hens out of their runs into a coop on wheels that can be dragged about. We put it in the meadow field behind the cottage garden with an electric net around it. The morning we move them is a bright, fierce one. When we let them out of their runs they dash into the dazzling green of the garden and split in all directions, scattering us after them. The last bird we catch is Harry. He is huge now: grandfather sized. He is good at the job of patriarch: in the evening he rounds up the hens before bed. He will not go in himself until he has clucked the last one up the ramp. If I go into the hut afterwards I always find him settled on the straw with the hens in boxes above him, his big white back like a puddle of moon in the half-light.

We worry about catching him. He may be stately and calm but he reaches above my knee and his spurs are bigger than my thumb. Slowly, we back him into a corner. Ben goes over to him and lowers both his hands onto the cockerel's back. Harry squats down, freezes, then allows himself to be picked up. He stays completely still, arching his neck around to watch Ben as he carries him with two hands and all his strength. Harry is too big to fit under his arm. When Ben sets him down on the ground Harry waits a few moments before standing tall and adjusting himself. We watch him as he watches us. It feels like something intimate has just happened. He had trusted us and we had trusted him. When I look at Ben his eyes are full and I know it is because of this big strong bird who could so easily have hurt him, but didn't.

Just over a year afterwards I notice that Harry isn't moving around so much. He is old by then, nearly eight. One morning I go to collect the eggs and see he is not with the hens. I look inside the coop and see him, eyes closed, his neck scooped around his body like a swan, his head resting on his white back. He has died in his sleep, on the floor of straw where he always lay.

The summer, which had begun to shine in spring, stretches on for weeks and weeks with no rain. Something strange happens in the meadow at the back of the cottage. I begin to see wildflowers I'd never seen there before. I learn that in a drought the shallow-rooted grasses die off first. When they do, they leave space. In the soil lurk the dormant seeds of wildflowers, some there for many decades, waiting for their chance to grow. With the field unploughed and unfertilized, and the dominant grasses finally making room, now is their time.

When it does rain I notice that, a few days afterwards, a mushroom ring always grows in the same part of the meadow field. They are caused by an underground fungus, which sprouts lots of small threads – mycelium – in a circular shape. A year later the mushrooms – the fruiting bodies of the fungus – grow up out of the ground, creating a ring. Even though the grasses are now long enough to hide them I can still make out their shape as the green around them darkens. These are St George's mushrooms, dirty white and umbrella-like, with a ribbed underside of hundreds of white gills that it is as satisfying to run my thumb along as a stick against railings. They appear in exactly the same place each year, usually near their namesake day around the end of April, although this time the drought has made them a month late. A friend tells me they are a sign of old pasture. I read about some that have been appearing in the same place for over a hundred years.

I like the idea of an ancient mushroom ring, growing once more in a pasture field that was once ploughed and cropped. I like the idea of a dormant seed finally able to grow after years of being buried. They teach us to bide our time and wait, for there will be a chance, if we keep looking for it, for us to grow and claim the space we are meant to.

The hay in the meadow field is cut and tedded – whisked and fluffed by a machine of the same name which helps to dry it. Farmers call it 'wuffling', and when I see it I find out why. Once, when I am in the garden and the day is still and calm, some of the hay lifts into a whirlwind, spinning on a thermal into a 'hay devil'. By the time I call the children to see it, it has gone. Later that evening Wilfred tries to find it again, running barefoot and half naked through the hay. His legs fly so fast his feet don't seem to touch the ground. I see how he is losing his baby body and taking on the shape of a boy. Now, though, flying through the meadow hay he does not look like a boy but a small deer, running and laughing and looking for the devil.

I notice that I am beginning to see the landscape around me differently. I try to read a field's topography for clues of what once lived on it in the way a surfer might read the ocean for rip tides, reefs, or breaks. I wonder what stories lie beneath banks, or ditches, or scoops or mounds in the earth. I learn that single trees in fields, cropped and harvested right up to their root balls, might show me where a hedgerow would once have been that joined the trees together, or where an old wood pasture once grew before it was cleared for agriculture.

I start to look at hedgerows differently, too. Where once I just would have noticed if a hedge was nice and tidy, and therefore thought the farmer a good one, now I see it otherwise. Towards the end of the summer a long series of hedges on farmland not far from us are given their annual top and trim. They are cut into short, neat oblongs less than hip high in a sweeping line up the field incline and beyond. The cut has taken with it any possible bird food for winter and there is little dense bush left for anything to make its home in. Great gaps already dash their way along its length where the hedge's top and sides have been cut but no gaps replanted. The branches at the hedge's bottom are woody and straggly. Eventually, I realize, the whole hedge will just slowly disappear.

There is no maintenance reason for the hedge's cut. It is metres away from the public footpath and separated from it by a ditch or – as some call them – a peasant trap, dug deep and wide enough to stop people walking on the margins. It has been cut out of habit and neatness.

Tidiness and competence have long been linked, particularly by my tribe: the middle classes. Messiness is encouraged only by the aristocracy or artists, especially so if you are both. My mother once told me about a neighbouring farmer who reputedly lined up his tractors with a ruler. They would be parked in the yard, gleaming clean and standing straight as a battalion. After all, they had just fought a war. When they were called upon these shining machines would roll out across the field wielding an armoury of weapons. Sprays and cutters went in to fight against any green that stepped out of line and sprouted where it should not have. But now I learn that if farmers cut each side of each hedge every year, the regrowth they remove is what produces the most tree and shrub flowers. Only climbing plants, like brambles or roses, produce good fruit crops in hedgerows cut every year. Every year a hedgerow is left uncut, says Natural England, it will gain two species of breeding bird. Flailing a hedge each year doesn't just lose the food for birds. Butterflies and moths will have laid their eggs on the regrowth. Autumn and winter hedge cutting is said to be one of the primary reasons why the brown hairstreak butterfly has become so rare. Others say the secret lies not in a fixed regime of cutting or not cutting, but the variety of it. Cutting hedges at different times in different ways in different years means that a host of creatures will choose them for their homes, depending on what they need and the food grown upon them.

Hedges, like people, hold the secrets to who they are in their roots. I have learned that if you take soil from underneath or alongside a hedge you will be able to see what kind of earth once lay in the field, before we went to war with it. A Cumbrian hedge will speak a different language, one whose words sound more Norse than English. The tree branches within it will say they have been lagged, not laid. Hundreds of miles away a Devon hedge may confide that it is far, far older than the parish church, underlain by banks built in the Bronze Age 4,000 years ago. Its neighbour might boast to be older still: a remnant of the original wood that once covered the field. A Dorset hedge might stay uncut so long that it will grow itself into a leafy green tunnel. When it is butchered back it will shiver with relief and silently reassure you that the following spring an explosion of primroses and light-loving wildflowers will grow again beneath it on the verge. In another five years the same hedge will have grown big enough to start to form its leafy tunnel again, and so the cycle goes on.

I look at the decapitated hedge and wonder what I would have seen a few years ago, before I was taught that, like people, I should not judge a hedge by its surface but instead by what lies beneath and within it. It makes me think of an old boyfriend who said he had finally figured me out. On first impression, it seemed like I had it all together. But when you looked a little closer you'd notice the flapping sole of my shoe, the fly of my jeans held up with a safety pin, the biro notes I'd written on my hand because I had no paper. From afar I looked in order but, once you knew the clues, you'd see I was anything but. Now I see these hedges trimming the Suffolk winter skyline in the same way. They look neat and precise, like someone has drawn them on with a ruler. But look a little closer, know what you are looking for, and you will see they are barely holding it together.

As I walk home I wonder why I am so hurt by the hedge being cut so hard. I have no right to it: no say over it, no claim over it. It and the land it grows in is not my land. I am just gifted access to it for it is part of the 180,000 miles of public rights of way that criss-cross England and Wales. I wonder if what I am feeling might go to the heart of the tension between those who farm the land and those who visit it. The former own it, but the latter feel an emotional claim to it. It is this which makes people talk of 'my park', 'my wood', 'my river', 'my hill'. They have laid memories in those places: they have been the stage for joy and grief with people both alive and dead. They find meaning in returning to them. When a place belongs to us in spirit we take its destruction personally. We may have no legal right to mourn a felled tree, a grubbed-out hedge or a stream that has been dredged and straightened. There may be reasons it has been changed: rivers must be sluiced of silt washed off the fields; coppicing woods and flailing hedges, which can look like destruction, keeps them alive. No one has any duty to explain the changes to us. But the shock of seeing it happen to a place we love can sometimes be strong enough to feel like a physical hurt.

As I see the landscape around me go through another cycle of harvest, I begin to suspect that we are designed to engage with the places around us, specifically with the natural world. That we too – like plants and trees – feel a desire to put roots out into the land and claim a space and connection for ourselves, even when the land is not ours. In return, we feel a little like the land has claimed us too. We don't need to have been born there to feel like this, nor to have known a place from childhood. I wonder sometimes whether this feeling is an echo of something once common to us all – long before industrialization, immigration, or being taken from our home against our will: a sense of deep connection to the natural world so that it might provide us with all we needed: food, shelter, life, understanding.

Later I am taught a word by a farmer's wife. *Hiraeth*. The Welsh say it is untranslatable. It means a sort of longing, a gravity-pull for home, or the idea of home, or the idea of a memory or feeling of a place. There is joy in it, in the sense that home is there somewhere, but also melancholy because we cannot reach it. It is not a real place. Sometimes I wonder if many of us in the city who feel restless are actually searching for this. Not so much a place to call home, but a feeling to call home. When I read about our growing disconnection from both the land and one another, I wonder whether we might be suffering a sort of communal *hiraeth*: a longing for something less superficial, something deeper, something rooted. For something that feels like home.

As I walk back towards the cottage I wonder whether I am now putting down roots. When I see it appear at the edge of the meadow field I am surprised by how much this landscape feels, absolutely and immediately, like home. And so I have my answer.

The new hedgerows in the pasture fields are planted in the early autumn, when the grass has stopped growing and is easier to clear. Afterwards there is little to look at except fenced-in saplings and bushes poking out of the top of rows of plastic tubes.

Farmer Martin, whose cows are still waiting for the organic conversion period to pass so they may graze this land with ours, has baled the hay in the meadow field and, one evening, I go and take the children to play on the giant golden cylinders, set at intervals across it. The air has the smell of late September, musky and rich. The ground feels old and tired, as though it is longing for rest. The swallows are hunting over the cut grasses. I wonder if we will see the owl too, but the boys yell and shout as I push them onto the top of the bales for them to slide down and the owl stays away.

Standing at the top of the meadow slope in the dusky light I notice how the young hedgerows completely change the feel of the landscape around me. Each of the three pasture fields now has a triangle carved off it: fat at one end, thin at the other. I try to imagine what the old hedgerows might have looked like a hundred years ago before they were pulled out. I think how cows, or sheep, or horses might have sheltered by them and browsed their leaves for food. I wonder whether people walked back from the village alongside them, winding through the hay stooks that would once have stood where these round bales now do. Whether children from the village picked blackberries, or sloes, or rosehips there. I think of the farm workers who would have cut those hedgerows back by hand to ensure they grew back full and bushy, and wonder whether they did so resentfully or proudly, teaching their children to do the same so they would know the skill when their turn came.

The new small triangular fields are now good only for grazing animals. We have turned out our three Red Poll cows into the one nearest the river, at the bottom of the slope I am standing on. They are different from the large black-and-white ones I see in other bright green fields. A cross between a Suffolk Dun – good for milk – and a Norfolk Red – good for meat – they were bred in the early nineteenth century as a dual-purpose cow. But in a world where only specialists are wanted they are now a rare breed. They are small and flat-backed and their coats glow chestnut in the evening sun. As I watch them graze I think how absolutely right these Suffolk cows look in this new small field. Somehow, in the early evening light, they seem to have made the meadow old again.

The nights draw in earlier each day until suddenly it is December and the year almost at a close. On the afternoon of the winter solstice – the shortest day and the longest night – I leave the children having tea, pull on a pair of boots and a coat and go outside into the garden alone in the winter dusk. I have spent the day cutting branches of evergreen and pulling great lengths of ivy from trees in the copse to hang around the cottage for Christmas. Tomorrow we will get a Christmas tree from the greengrocer, who will let the children choose the fullest, most fat-bottomed one. I think how Christmas here feels different from a city Christmas. There is none of the promise of lights and parties and streets full of shoppers. It feels deeper: more ritualistic. I have read that some trace our Christmas feast back to Saturnalia: an ancient Roman solstice celebration dedicated to Saturn, the god of agriculture and time. As I walk over the darkening fields I can see why agriculture and time might have needed their own same god.

The winter air is cold and sharp. It is the kind of night I have come to love. Later I will be able to see stars from one side of the horizon to the other, sometimes clear enough to make out galaxies that make the sky look as though someone has swept a silver paintbrush across it. Now it is filled with the sounds of rooks barking to one another as they roost in the branches of the trees at the edge of the field.

I walk to the end of the garden and lean against the fence as the trees become outlines of black against the dusk. I wonder how this view will change when the hedgerows and trees we are planting have grown taller than I am now. I think how I will probably not get to see their full height in my lifetime and what a strange feeling this is for someone more used to immediate gratification. Maybe one day someone else will walk these fields and look at all we have planted here, and assume not only that these hedgerows, trees, pasture fields and scrub have always been here but that they always will be. Maybe that person will also be a mother with two small children, wondering where life is taking her. I realize this means I will be linked to this person even if I am long dead, and we share no blood, and they never know my name nor how I helped shape the land they are looking at.

I realize both that I have never had this feeling before and also that farmers must have it all the time. They read the landscape like a map, except that rather than coordinates they mark it with memories. Cottages lived in by those who worked on the land and shared their language. Barns built and barns pulled down. Fields bought and fields sold. I have met many farmers who are aware they are a thread in a long personal and local history. There is one who affected me more than most. Tom is a dairy farmer living in the north-west of England. He is young and modern and ambitious. He is a dairy farmer because his father was, and his father's father too. He loves the feeling of watching cows graze outside and so this is how he keeps his, even though he could make more money by keeping them in barns like his neighbours do, for he is farming in a world which no longer really values the milk he makes. This twin tension – of wanting to produce something that many consume but few appreciate – has seen him do things he did not want to.

He bore the human cost of our choices, which so often seem to stay unstudied in our effort to uncover the ecological and environmental ones. He could have hardened himself against it as others have and said, 'It's just the way it is.' He could have found another life. But he didn't. Instead, he changed his.

*I'd just like to look in the mirror and think I'm doing*
*the best things that I can.*
*I don't have — my knowledge of farming is small — but*
*I'm doing what I think is right, basically.*
*Because if I'm going to do this for the rest of my life, if*
*it's going to be fairly all-consuming in lots of ways, then*
*I feel like I've got to be doing it so I feel good about it.*

Chris, farmer

# 6. Tom

The year 2010 begins with the worst snowfall in two decades. The British media call it *The Big Freeze*, which sounds like a Disney character and is far friendlier a name than it deserves. In the barn alongside the dairy parlour, Tom and his family help deliver new-born calves then tried to save them from the shivering cruelty of pneumonia. Despite their efforts, many of them die. Commentators will warn everyone that extreme weather events like this will soon become the new normal. We must, they say, begin to learn how to live alongside them. Floods, gales, droughts, storms: the weather of biblical punishments.

On a morning in February, Tom's alarm shakes him from his bed at five o'clock to begin milking. He dresses in quiet, cold darkness, pulling blue overalls on over his jeans and sweatshirt, careful not to wake his parents, with whom he shares the small farmhouse. He doesn't, therefore, turn on the radio but, had he done so, he might have heard news of another foreshadowing. An epidemic flu, jumping from pigs to people and killing them in the process, is finally beginning to subside. Precautions are being relaxed. The people of Britain find themselves accused of being a 'nation of hypochondriac drama queens'.[1] The former head of the Council of

Europe's health committee accuses the World Health Organization of having 'faked' a pandemic.[2] Conspiracy theories about governments' response and the company that created the vaccine abound. In the end, some reports will conclude that up to 575,400 people die of swine flu, but most people's lives will stay unchanged. Few will feel it as a warning of what's to come in just a decade's time. Only some forecasters will later turn and cry: *See: we told you so; we knew that if you kept on as you were, something big would come to bite you.*[3]

But Tom doesn't listen to the news that morning. Instead, he presses his feet into thick rubber boots in the small brick porch in silence, then opens the front door and walks across the yard to the barn where the cows are waiting for him, their coats and breath steaming in the darkness.

There is a radio in the milking parlour too. Once Tom has climbed down the steel steps into the parlour's belly and strapped on his rubber apron, he could have listened to the news then. While he opened the clanking metal barriers, letting the first group of bawling cows into the parlour, he could have listened to the debate raging about too much money being spent on preparing for pandemics that never properly materialize.

But Tom doesn't listen to the news when he is milking. Instead he listens to music. He turns the radio up loud so he can hear it over the machines and the cows calling for their turn. That morning he is joined by Katy Perry, Cheryl Cole, Lady Gaga, The Black Eyed Peas.

Their voices echo around the steel trench as Tom moves from one cow to another. The machines sing a bass line — *voot, voot, voot* — as they pull fresh milk from the cows along clear plastic tubes. Some find this place claustrophobic: standing below ground level surrounded by beasts and hooves and shitty, flicking tails. Tom doesn't. He just hopes for an end to months of dark winter mornings and fingers red and rough with cold. He cleans his hands in the small sink at the end of the parlour, wipes them on a sheet from a roll of blue paper towel, and just gets on with it. He sprays teats. Wipes them with a paper cloth. Attaches the clusters, hoovering their way onto each udder's teat. One, two, three, four. Avoid the piss, avoid the shit, avoid the kicks. Tom moves from one cow to another, all the while trying to block his mind against the rest of the day, and what's to come.

Afterwards, Tom mixes up milk powder to feed the young calves. The heifers will be kept to join the herd, but this breed of cow is good only for producing milk and the bull calves will never grow fat enough to be sold to raise for meat. Were Tom to drive them to auction after weeks of feeding them, keeping them well, paying to register and tag them, he'd be lucky if they fetched more than £10 each. So now, with every single penny squeezed, the farm has little choice. The bull calves have to go.

The milk will be collected by a large tanker later that day: a shining barrel of refrigerated steel that has to reverse up the farm's long drive so it doesn't get stuck trying to turn round in the farmyard. For every litre of milk the farm sells to the processor who owns the large tanker Tom's farm will be paid, on average, 24.67 pence.[4] Fourteen pence a pint. It's not the worst price they've ever got. A few years earlier they earned even less. The problem is it costs the farm 36.6 pence a litre to keep and feed and nurse and house and water and care for the cows who make it.

Three hours after he has begun, Tom finishes milking. He cleans all the equipment, hoses the floor down, switches the music off and walks back across the farmyard for breakfast. The brass knob on the farmhouse door is worn dull by working hands, but none who touch it daily are there when he enters the empty kitchen. While he has been milking they have all begun their day. He can't decide if he is relieved or disappointed.

Opposite the front door is a line of cupboards underneath the kitchen sink and, above the sink, a window that looks over the fields beyond. Along the adjacent wall sits an old dresser, filled with plates and teacups hung from small hooks screwed into the wood. The wooden kitchen table is pocked with dips and whorls where three generations of elbows have rested. Alongside the front door, next to the window that overlooks the farmyard, sits an oak chair with curved arms. It has a thin, worn cushion of deep berry red with the seat bones of Tom's father imprinted on it. This is his chair, and no one will sit in it with him in the room.

His mother has left him bacon in the pan and the kettle warm. He makes toast to eat with it and sits at the kitchen table. Usually, at breakfast, they will all work out their jobs for the day. If his dad is still there they will divide what's to do between them, depending on how kind or malevolent the weather feels. That morning, though, Tom sits alone. He knows what his job is. There's a pull in the pit of his stomach that tells him so.

At the end of the table is a newspaper, which, had he read it, would have told Tom that, as well as snowstorms crippling the country and a killer pandemic sweeping the globe, there has also been another rise in unemployment, with more than 2.5 million people out of work.

Tom finishes his breakfast without reading the paper. He is not thinking about global pandemics or milk prices or unemployment figures. Instead, he is trying hard to empty his head of any thoughts at all. He walks over to the sink and puts his empty plate into it. He had hoped that the dirty breakfast things might still be out. Then he could claim to be busy as he cleaned them up – maybe even unavailable – but, of course, their small kitchen is spotless. Other farmhouse kitchens are usually never really clean – there are always bits of baler twine or tools or folded-up feed bags or a coat drying over a chair or boots kicked off across the floor or just clumps and clumps of never-ending mud – but then other farmhouse kitchens haven't met his mother. Here, boots and coats must stay in the porch, the floor brushed, the countertops wiped, the breakfast dishes washed and the chairs around the old wooden table all pushed in by her. Tom stands at the clean kitchen sink, staring at the tiles on the wall. They are cream with a picture of a small brown hay stook on them: an emblem of farming's past. The picture is so familiar that he usually no longer really sees it, but now he wonders at the history of these hay stooks, and the history that led his family to this land, and the history that means his day is overshadowed by a leaden feeling of dread.

The small brick house that Tom lives in with his parents is joined to the road by a very long, mostly straight path. Either side lie sprawling green fields, trimmed with hawthorn hedges, tall and thick enough still to be stock proof. Around them the land is flat and wide. A little distance away from the farmhouse stands the barn the cows live in for the winter months, to protect them from the weather and ensure they do not churn the ground that feeds them into uselessness. Behind them in the distance lie the hills, rolling away in enormous emerald waves.

Tom is here because his grandfather came down from those hills in a horse and cart to look for work, a story that has been absorbed like folklore into his family. Finally, his grandfather was able to rent this farm from the old landowner. When a change of taxes in the 1980s made the ownership of land less appealing, the landowner – like others who owned the land for money, not love – sold it to his dad. A few hundred acres: a farm of their own.

Once, when he was younger, Tom's father told him that the hill sloping up and away to the west was known as Peat Mountain. Back when he was young, his dad had said, the men would dig up peat from the hill, stacking it at the top of the fields. Spadefuls of dark earth were piled into bricks. People would come to cut or collect them, then burn them in their fires and stoves in winter. His dad told him that the peat was made from rotting vegetation or organic matter, but not that it was the most efficient carbon sink on the planet, because no one really talked about carbon sinks back then. Few knew, or thought about, peatland covering just 3 per cent of the world's land's surface but storing one third of its soil carbon,[5] or that it took a thousand years to produce just one metre of it. It was traditional for folks to come and cut it, and peat was cheap and burned and burned, longer than logs would, and there seemed to be so much of it. No one really knew or understood, and people were cold, and what else did anyone really expect?

Underneath the fields around them there is peat soil too. This, together with the rain, which here falls and falls while other parts of the country stay dry, means that they can grow grass for fun. This is poor land for cropping, perfect land for grazing. When others are forced to buy in feed because their grass is parched by these new spring droughts, Tom's farm can still get two cuts of grass off just one field. Unless, of course, another of the old drains that stop this land turning into bog decides to burst, taking not just the grass but the Christmas money too.

As soon as the rain stops falling and the sun returns, the fields around the farmhouse will be covered in the farm's cows. There have always been black-and-white cows on this farm for as long as his family have farmed it. Tom's grandfather had bred Friesians, small black-and-whites that had – he always said – nice temperaments but lots of personality. Although his grandfather could not have been called a sentimental man, he knew each one of them. The skittish one, the bolshie one, the gentle one. On his old farm there were just twenty cows, and they spent the winter tied up in a barn with a rope around their neck, as most did back then. The stalls they stood in would never have housed a modern cow, not now everyone thought bigger was the same as better. When the winter was over, his grandfather turned the cows out into the fields. Tom would stand with him by the gate and together they would watch the cows buck and jump with joy, galloping along each side of the field in turn, measuring out their boundary. 'Daft buggers,' his grandfather would mutter warmly. Tom would look up at the man who, by taking to the land, had brought the generations of those who followed him onto it too. And so he was taught as indisputable fact that this life was now in his blood and bones as well.

But in the 1990s, when Tom was too young to have a view, everyone began to want a new kind of black-and-white cow. Holsteins: the racehorse of cows, super-cows, bred only to produce milk, the beasts towered over their black-and-white Friesian cousins. They had wide flanks with jutting hip bones and long legs and udders that could bag up so big it was sometimes hard for them to walk straight. Their value lay only in the volume of the milk that seemed to flow endlessly out of them.

These new bovine machines could produce double the amount of milk the old cows could, more if they were milked three times a day. Now there are half the number of cows in the UK that there were when Tom was born, and yet they produce nearly twice the amount of milk.[6] To produce all this milk these new kinds of cow cannot be fed on grass alone. Tom would help his father unload the supplementary feed bags from the delivery truck and watch his brow furrow at the invoice, and wonder at how mad it was that we have made a cow that cannot live off grass. That we have made a cow that needs so much food to make its milk that, were they to keep it in the field and continue to milk it the same way, it might even starve to death.

It is not just the additional feed. The week before that cold February morning, Tom and his father had been walking the cows from the barn into the parlour for afternoon milking when one had slipped over on the wet concrete floor of the yard. His dad had spent money putting down a special pathway from the field to the parlour that was supposed to be gentle on the Holsteins' feet, too soft and delicate to bear their great weight and with hooves that had to be regularly trimmed and monitored for signs of the lameness that plagued them. They had to take out another loan to pay for the special pathway that was also supposed to stop the cows slipping when it rained, which it did, but when they were on the concrete, little could help them.

Tom was standing towards the back of the herd when it happened. The cow, one of the last to come into the yard, went down to the ground right next to him. She was a good cow, friendly and gentle. Young too, only two years old and not long calved. Now her long back legs split either side of her body and she panicked, scrabbling on the concrete, her eyes wide with terror. It is, Tom thinks, the most pathetic thing to watch Holsteins try to stand once they're down. Udder heavy, sometimes they tried to get up like a dog, front legs first. He could tell that the cow had popped a joint and cursed himself for letting it happen, even though he knew there was little he could have done. He hoped he might be able to lean on her and click the joint back in then try and get her to her feet. If he could she'd be all right. Otherwise that fall would cost them another cow. In the end, despite his efforts, it did.

The night after the cow fell, Tom and his father sat at the kitchen table and worked out that, over the last year, they had lost forty others. Their hearts had given out, or their stomachs had become twisted and could not be stapled straight, or they had fallen and could not get back up again. Around this farm are eight others, all dairies like theirs. By the end of the following decade only one other farmer will let their cows out to graze. Tom understands why: the slipping accidents; the lameness; the increased milk and income from being able to feed cows a diet as bespoke as an athlete training for a marathon. It makes sense. When the world is upside down.

Somehow it doesn't strike Tom as unusual that not once does he or his father or, when he was alive, his grandfather, ever countenance doing the same. Cows belong on grass.[7] Tom has absorbed this as a fundamental truth. The sound of cows tearing grass with curled tongues, the sight of them in the field, grazing with their heads down or lying in the shade of the hedgerow, are part of his language and his understanding of the world. Letting the cows out to graze when spring arrives is one of the best days of the year, a kind of ritual, a ceremony of the season of darkness passing into light. His grandfather and his father liked the way it made them feel as much as Tom likes the way it makes him feel. And so an unspoken agreement between these three generations of men has been struck, which means their cows go out to graze grass, no matter what it costs them.

But it is costing them. The only way they will survive this way of farming, Tom thinks, is to get bigger. So they have bought more cows, but more cows means more winter barns and more money spent on straw, feed, fertilizer, diesel, seed, equipment. It seems like everything is going up in price except milk. His grandfather's herd was a third of the size of theirs yet even with their new outsize cows and more milk than ever, everything is tight. There is no fat: not on the cows and not on the balance books.

At night the figures of it go around Tom's head as he listens to his parents in the kitchen, trying to work out what to do. His mother washing up. His father going through paperwork in his chair by the window. They talk, swirling Tom's future up with theirs. His dad growls about the supermarkets and how they are screwing him. The new contract they want him to sign is even worse than the last. They want the farm to detail all their production costs: costs of the barns, the feed, the vet bills, the farmhand who helps with milking, but he knows what will happen if he does. The supermarket will come back and tell them where they can cut those costs even further so they can pay him even less. No, he's not having it, says his dad. They've been screwed long enough. They need to try for something better. A company where they have some control. He's heard of a contract, says Tom's dad, that others are joining. A kind of co-operative. You have to buy into it: 1p a litre for five years. Tom hears his mum whistle through her teeth, hands in the sink, not turning round. But this way you have a say, his dad comes back, you have a voice. A small one, and you still must take the price they give you, but still. It's a voice. Most of his father's friends have already refused to join this new co-operative. Too tight for their own good; typical, his dad says. Besides, what have they got to lose? How much longer can they keep going as they are?

The discussions usually end in silence. Tom hears plates being put away. Cupboard doors banged. From the way the cutlery is dropped into its drawer he knows his mother is upset. He can picture his dad's face watching her. His face is becoming grey and thin in a way it is not in Tom's mind. It will be years before they discover the cancer. His mother will be sure it is the stress which causes it. She reads something once that says this, but no doctor will confirm her suspicion: that this job – no matter how he loves it, no matter how much it is part of his being – that this job might be the thing that kills him.

On that cold February morning, Tom stands in the kitchen staring at the only kind of hay stooks he will ever see, and feels dread settle in his gut. Then he hears it. The sound of a truck's tyres on the long drive that ribbons up from the road, past the two empty grazing pastures. He hears the truck door slamming. A pause. Voices in the yard. His father calling his name. He doesn't answer. He knows his dad hates this business just as much as he does, but it has to be done. There's not much of a choice. Kill the bull calves, or kill the farm.

Tom listens as his dad's boots walk towards the door, his gait and tread as familiar to Tom as his face. He hears the front-door handle turning. He knows his dad is standing in the door frame.

'Tom? Knackerman's 'ere. Get out 'ere, lad, and help 'im get rid of them bull calves.'[8]

A month later, Tom is sitting at a pub table with five strangers on the edge of town, near enough for most of the people in it to be from the city. After-work drinks; date night; suits and heels.

He is meeting a friend, who, unlike many of the children he went to school with, left the countryside and got a job in town. Now Steve wears a black suit and a white shirt and sits at a desk and, even though Tom knows almost everything about him, when he walked into the pub that evening and saw Steve sitting there, he felt a pang of self-consciousness about his own clothes. Tom looks at the others around the table. There is Steve's girlfriend, Stace, and her friend from work, Cassie, who has green eyes and auburn hair and pale skin with dots of freckles that look like they've been flicked from the end of a paintbrush. Now Tom catches her eye and wills the flush he feels creeping up his neck to go away.

One of Steve's workmates asks him what he does, and, when he replies, says in response, 'Oh, I know a farmer!' like it's the same as knowing someone off the telly. Tom has discovered people often say this, but their farmer usually turns out to be someone they have met a few times, or who they don't really see any more. 'That's because we're always working,' he jokes. It's partly true. Their days are long and the nearest pub is often a can't-drink drive away. But he also knows most farmers prefer each other's company so that they can talk in a language few others understand, not having to explain or justify themselves to those who will eat what they make, but judge how they make it. His daily life is so different from these people's. He'd once met someone in the army who'd said the same. No outsider really gets it: they can never know what it is actually like.

He knows it won't be long before the questions come. He doesn't mind; sometimes he quite likes it, really. The men always ask whether or not he gets to drive a tractor (yes) and combine harvester (no). But it is usually the women who ask him the question that surprises him the most. Stace does it now, leaning over across the sticky pub table, her face pink with heat and drink.

'So, like, does a cow have to give birth to, you know, make the milk?' The boys all laugh and mock her but Tom sees how they turn and listen to his answer.

'Yes,' he says. 'They'll be put to the bull every year until they no longer produce enough milk. Then they'll be culled.' By which he means killed. By which he means the cow will be eaten, by either a human or an animal (although, he'll think but not say, *What's the difference, really?*).

On their farm they have now started to buy in expensive sexed semen and, if the cow is a good one, she will be impregnated with it: an immaculate conception with a long plastic glove and thin metal tube.[9] This will nearly always guarantee a heifer calf. Then, unexpectedly, he suddenly pictures standing in the cold of the calf pen, helping the knackerman to shoot one healthy bull calf after another. He has no problem putting an animal out of pain or sending it into the food chain at the end of its useful life. Tom is a farmer. To him that means he farms something, be it the land or an animal. It involves taking something off something else for food. What, he sometimes wonders, do people think it means? But what he must do with the bull calves is different. It feels different. And, each time he does it, it leaves a sort of scar on him.

This time, though, Stace doesn't think to ask what happens to the bull calves. Instead she leans across and says to her friend: 'Here, Cassie, didn't you grow up in the country?' When he looks at her, Tom notices Cassie's cheeks dot with pink, right in the middle.

'Pint, Tom?' Steve stands up, heading for the bar.

'Sure. Thanks.'

That day, Tom had read an article that said a pint of

beer now costs more than a chicken. This pint, that Steve will put down on the soggy beer mat, is valued at a price greater than an animal that will feed a whole family. An animal raised and fed and housed by a farmer who will, Tom knows well enough, get a fraction of the cheaper-than-beer price it's sold for. Tom knows some people think it's the farmers personally who benefit from government subsidies. That somehow the public's tax money is going straight into their pockets alongside all the profit. But the public's tax money is getting the public cheap food, that's all. No food is priced to show its real cost. Not the monetary cost that it takes to produce it. Not the environmental cost. Not the animal cost. And not the human cost either.

'So, you're a farmer then . . .'

Tom looks around. Everyone else is talking to each other. He looks back at Cassie.

'Yeah,' he tells her.

'That's cool.' Cassie reaches into her handbag and pulls out a partly drunk plastic water bottle, slightly scrunched. She flips the sports lid off and takes a swig of it but squeezes too hard and the water leaks down her chin and onto her outfit. She cries out and pulls back, the legs of her navy suit trousers darkening with the spill. On instinct, Tom reaches into his pocket, takes the handkerchief he always carries – as his dad still does, as his grandfather always did – and gives it to her.

'It's clean,' he says, lying.

Cassie dabs the handkerchief at her legs, tossing the nearly empty bottle onto the table. Tom knows if the bottle was filled with milk, not water, it would have cost her less. Milk is now cheaper than bottled water. He knew that without having to read about it.

He doesn't mind tonight, though, because the water lets him help her, and because of that she gives him her phone number and, later on, her heart. When they kiss outside the pub that evening he doesn't talk about beer costing more than meat and milk less than water because he is falling in love. And anyway, he's just not that sort of guy. Besides, that morning a supermarket delivery van drove up the farm drive with his family's weekly shop. Food is too cheap and because of that they cannot afford to buy it anywhere else.

Tom falls for Yolanda the day she goes into the field.

She arrived on a summer's afternoon along with a dozen other cows, all the way from Holland. They had spent the night in the barn and, that morning after milking, Tom went into the barn, opened the gates and began to walk the new cows out and around to the edge of the field.

It is June 2011, just over a year from the day that Tom stood staring out of the kitchen window wondering about the history that had led him here, and willing the world to right itself. The sun is strong already and the green of the field is bright against the sky. In other parts of the country cows graze in fields with patches of grass brown with drought. In their field though, the grass is plentiful, and the cows tear it from the ground with sandpaper tongues or lie in the shade of the hawthorn hedge.

As he walks alongside these new cows Tom is struck by how unused he is to their shape. Smaller, hairier, meatier, their flanks are covered in flesh and muscle instead of skin and bone. These new cows trot where the Holsteins once loped; are skittish where the Holsteins were dopey. They have French blood; Scandinavian blood; Channel Island blood: each of these new cows is a multinational mixture.

The one at the front, leading the herd, will become Yolanda. Tom walks alongside her and notices the curling white forelock that juts vertically between her ears. She seems friendly enough and so he rests his hand on her back as they walk, interested in how compact she seems compared to the Holstein giants. When she gets to the edge of the barn, opposite the open gate to the field, Yolanda stops and stares at the patchwork of green in front of her. The other cows halt obediently behind her. Tom watches her. This cow – these cows – are a risk; a gamble. But they are also the beginning of something. Change.

'Get up then, lass,' he says, slapping her on her rear flank.

He need not have. She does not need the encouragement. She has not stopped out of fear or laziness: she has stopped out of shock, because this is the first time she has seen grass. When this subsides Yolanda trots then runs at the field, the rest of the cows following behind. When her hooves hit grass she kicks and bucks and jumps like a bronco, bouncing across the green. Tom leans on the fence as the rest of the cows canter past him and follow their leader around the field's boundary. Tom watches Yolanda and thinks, *Well then, life is changing for both of us, isn't it?*

The field is large and Tom assumes the cow will soon stop kicking and running and get to eating. But she doesn't. Yolanda keeps going. Sensing trouble, Tom climbs over the gate into the field. The field is bordered by a big thick hedge that no cow has ever got over. Yolanda, though, runs straight at it. Her big bovine head ploughs in one side and out the other, where it stays. Her body, too large to follow, struggles for a bit before giving up. Then she simply stands there, gazing across the fields that lie on the other side of the paddock, spellbound.

*Shit*, he thinks, *that'll take some mending.* He goes over to her flank – *Go on, girl* – and with his hands tries to push her body back but she locks her knees and refuses to move. He walks all the way along the length of the hedge and around again to her head, now framed with a frill of green. Yolanda sees him coming but doesn't budge. Tom stops in front of her, wondering what to do. Tangled in her forelock are a few of the needle-sharp spikes of the hedge, long enough to have blinded her. The white flowers that had speckled it with prettiness in spring have mostly gone over, but there are still a few left. One has rested in the crease of the cow's ear as though she is a picture in a children's book. Tom looks at Yolanda and wonders what to do, as she looks back at him with huge brown eyes. Then she opens her mouth and bellows and bellows into the valley and on to the green hill beyond it in sheer and utter joy.

'Tom! What the bloody 'ell is this cow doing in the hedge?' His dad's bodiless shout rings out into the blue of the morning.

'She's got herself stuck, Dad! Good and proper stuck.'

'What the . . . ?'

After a moment his dad comes round the corner of the hedge, putting his hand up to shield his eyes as the summer sun hits his face. He walks over to Tom and they both stand staring at the cow. She stares peacefully back at the two men. When Tom turns to look at his dad he realizes that he is standing exactly like him.

'Naw . . .' his dad says. 'She's no' moving.'

Tom grins . . . 'Yer mean, she's not mooooving.'

His dad snorts and slaps Tom across the stomach with the back of his hand, then lets out a bark of laughter. Then Tom laughs, in sheer delight at hearing him, and soon they are heaving with it, bent double with it, helpless with it.

'She's udderly stuck.'

'She's made a bulls-up of that there hedge.'

'She'd butter get outta there.'

It is not the jokes. It is everything. It is feeling like it might be OK. That the farm and their legacy and their future might be OK, or maybe it won't be, but at least they have changed something. At least they have tried. It is ten years' worth of tension expelled into a summer morning while a brown-and-white cow with a forelock stands and watches them laugh and laugh and laugh.

They try to pull her out. Tom puts his large strong hands either side of her head and pushes her backwards as his dad shoves her rear, but she just looks at him and does not budge. She doesn't want to move and no one is going to make her. So they leave her to it. Sure enough, by lunchtime she is grazing alongside the others, a hawthorn flower tucked behind her ear.

Eventually, over time, Tom becomes used to seeing the fields without black-and-white cows in them. He learns a lot, but the main lesson is to ignore everyone else. These cows are not hidden away in a barn but are instead outdoors for everyone to comment on. That is the thing about farming: everyone can find out all about your mistakes just by looking over the hedge.

His neighbours tease him about the cows in the pub. 'You still breeding them mongrels . . .?' they holler across the bar when they see him. 'What are you doing with those ugly cows, those peasant cows,' they say. It's not just mockery; it's also suspicion, as though he's hit upon a secret he isn't sharing. And he has. These new cows, built for strength and hardiness, produce less milk than their predecessors but they rarely slip or drop dead and the vet rarely needs to call any more. Their feet are black and strong. They live for nearly twice as long as the old cows who were worn out after just four or five years of milking. Tom knows some dairies that consider a cow old at seven. In his herd they often live to sixteen.

Most importantly, the fleshy flanks of these new cows mean they are as good for meat as they are for milk. Now he drives the bull calves from their pens to the auction mart to sell them, where they make £150, £250,

£350 each. Never again does Tom stand in the kitchen of the farmhouse trying not to hear the sound of the knackerman's van in the yard.

The milk contract they are now under is the one his dad wanted to join. There had been many lean years when they all worried it had been a mistake, and his dad grew greyer, and his mum smiled less, but now they are getting one of the top prices in the market. More than that, they are paid extra to keep the cows happy and healthy. There is not a lot of money but there is enough; enough for Tom not to have to do things he doesn't want to.

In the fields outside the kitchen window the grass has been mixed with clover and, inside the drawer in the kitchen dresser, are leaflets listing the herbs Tom is thinking about experimenting with and growing in his fields as well: chicory, ribgrass, forage herb, trefoil, meadow fescue, timothy, lucerne, festulolium, which will anchor the soil into place more deeply than the ryegrass in his fields can and give the cows a better balance of nutrients. Tom wonders what it will be like to see the fields speckled with colours from these plants rather than the squares of bright green he is so used to. He doesn't realize that, had his grandfather still been alive, he would have recognized all of these plants growing in the fields he rode past in his horse and cart on his way to become a farmer.

There are other changes too. In a shed opposite the cow barn sits a large shining tank, inside which is a machine that turns the cows' slurry into energy. Tom reckons it will pay for itself after five years. It already powers the majority of the farm and the excess electricity is sold back to the National Grid. If Tom's projections work out right then over the next twenty years, his cows' muck might end up making the farm more money than their milk. For while Tom's world has begun to right itself, the rest of the world is still upside down.

All around them, though, smaller dairy farms have been selling up. A few months earlier Tom had seen one of the farmers in the local post office. He had sold his farm last year and moved to France with his wife. His kids had grown up and didn't want it: why would they? Now he was back to sort some paperwork out. They exchanged greetings and talked, skirted around it for a bit until Tom asked him straight out how it was, leaving the farm.

'Best thing I ever did; wish I'd done it sooner,' the man said, tipping up his chin.

He looked happy and Tom almost believed him. Then he opened the door of the post office into a spring day that rang with birdsong and sunshine, every hedgerow and field on the cusp of bursting into bloom and, lit up by it, turned to shout goodbye to see the man – just a man now, no longer a farmer – watching him go, with an expression that made him realize otherwise.

It is summer 2020. Tom stands rinsing his breakfast plates at the kitchen sink after milking. It is his first shift for days, now he and his father share it with two others. The news is still too bleak to listen to: a global pandemic sweeping the world; extreme weather events; an increase in unemployment and redundancies. That morning Tom had turned the music up extra loud. Spray; wipe; suckers; dip. Avoid the kicks, avoid the piss, avoid the shit.

The front door of the farmhouse is open to the summer sunshine and, from where Tom stands, he can hear the sounds of two calves born the day before. They calve in the summer now to beat the cruelty of pneumonia. One bull, one heifer: both welcomed. His father had watched the two calves being born, although it will be the final time he does so, for his cancer will come back to claim him later that year. He is nearly lost alone in a hospital bed, surrounded by hazmat suits and plastic visors in the midst of another pandemic panic, this time one that everyone will remember. But then Tom's mother storms the ward and carries him home so he can leave this world from the one he loved: beside her, in the farmhouse, on his farm.

Later that summer's day I will visit Tom for the first time. He will take me out to the field to see Yolanda, who is now over ten years old. He cannot get her pregnant any more and not a drop of milk has come from her for years. She is one of the first cows to come over to us, slowly, for she has arthritis in her joints and has grown fat on her back from treats from the milk hands. Yolanda will not, Tom vows to me, leave this farm. She will die here when her time is up. On this he is resolute. She was the beginning. She was the one who helped change it all.

The sun is bright and I think how the scene I am

walking through looks like a postcard or the picture on a label, gummed onto the side of a milk container. It is the kind of image that someone in a city supermarket sees: blue sky, green grass, grazing cows. The kind a mother might look out for as she stands in the aisle, her toddler sitting in the orange plastic seat of the trolley, his fat legs banging against the metal bars as he grizzles. She glances at the label, then at the price of the milk, then at her watch and reaches for the cheapest one. I've done it. I've been that mother. I hadn't understood the cost of what I was doing. I hadn't understood what I was asking farmers like Tom to do through the choice I was making because I never properly understood the consequences of that choice.

That day I drive home through valleys of pure and perfect green. I think of Tom – unsentimental, pragmatic, tough – and how he had described hiding from the knackerman so he would not have to help kill the bull calves. I felt my heart hurting not for the poor dead calves but for him. I think how aware we now are of the ecological costs of farming. We are told that agriculture contributes 10 per cent of all UK greenhouse gas emissions.[10] That we are in a biodiversity crisis largely contributed to by fossil-fuel farming's use of chemicals and disregard for nature. That there are no rivers or streams left in the whole country now deemed safe enough to swim in – not one – and that while others play their part in this, agricultural runoff is still the leading cause.[11] But whilst we are all made aware of the price paid by the natural and animal world for the way our food is made, there has been a human cost paid too. Often it has been a deep and painful one which feeds directly into the statistic that means more than one British agricultural worker a week takes their own life.

I read and hear so many words of blame about farming but far fewer of responsibility. As I think back over the choices I have made – and still make – I wonder if it is not just separation from the land that has enabled this, but a separation from our farmers. Now we find ourselves on the brink of remodelling what being a farmer means. Tom has shown me that, if there is to be reform, it is not farmers who should bear the burden of change, but all of us.

*I think a farmer is someone who takes a natural resource and adds value to it without destroying the long-term quality of the natural resource. You know what I'm trying to do is put more back into this place than I take out — because it's had more taken out than put back — and get it as good as I possibly can.*

Andrew, farmer

# 7. Year Three – A Local Acre

I drive through wintery lanes on rounds of domestic duties and notice two things. That the brown branches and verges and hedgerows are finally starting to turn green, as though a watercolour brush has swirled in their outlines with spring. And that hedges and trees are being planted: everywhere. At each turn I see miles of thin plastic tubes, their bases surrounded by the bleached yellow of grasses killed off by the glyphosate used to clear the ground beforehand.

On the vast field across the valley from our pasture meadow, another newly planted hedge winds along its long roadside edge, altering the horizon. I find out that the landowner saw our new hedgerows go in and asked Ben how she might go about doing the same. Change, it seems, is catching.

As I drive, my children yell out from their seats in the back the names of the machines that pass us.

MASSEY FERGUSON!

NEW HOLLAND!

JOHN DEERE!

SPRAYER!

SEED DRILL!

Friends and relatives buy them tractors and books about tractors, and it is mostly from these that I start to identify the metal juggernauts that thunder past me. On the rare occasion I am driving alone and pass one barrelling its way down the long Roman road near us I catch myself thinking, 'Ah, a cultivator. The boys would have loved that.'

One day we line up the toy tractors from Ben's 1980s childhood against our children's new ones. They look so tiny next to them that we assume the toy makers must have changed the scale. When Ben looks it up we find we are wrong. It's not the model scale that has doubled; it's the size of the machines.

Sometimes it feels like farm machinery has become linked to status in the same way a big house or flashy car has. After all, this is an industry where sales of trade magazines reportedly increase by a measurable percentage when a new tractor is pictured on the cover. I've seen grown men walk up to tractors as though they are a deity – which, I suppose, they sort of are: a kind of mechanical god, able to plough a forty-acre field in an afternoon when once, when horsepower did the work, it would have taken sixteen ten-hour days. These machines thrust through fields with the kind of size and power that make their riders feel Cock of the North, South, East or West. They have self-deflating tyres,[1] air-conditioning, heated seats, two cab screens, Bluetooth speakers, a fridge and the shine of the future. Some cost more than the average British home.[2]

Their link to the past has not been entirely eradicated. Their power is still measured in the horses that would once have pulled the plough; the fields they drive across are still measured in acres, a word going back to the Middle Ages which describes the amount of land that could be worked by one man, behind one ox, in one day. Still, standing outside looking in, it is hard not to want to join a link between these monster machines and a certain kind of masculinity that wants everything to be bigger and more powerful.

The weight of an average 150hp tractor has more than doubled in the last thirty years.[3] While egos are puffed, the soil is crushed. It is not always the topsoil that is compacted, but soil at greater depth, and no matter how well machinery weight is spread with clever tyres, the sheer mass of metal sitting on top of them has an impact. Compression affects the physical, chemical and biological processes taking place in the soil. It causes decreased crop production, root penetration, nutrient and water uptake and waterlogged soils. Some have called it the 'costliest and most serious environmental problem caused by conventional agriculture', which seems quite a high price to pay for a tractor turn-on.[4]

Farming, the oldest industry in the world, has always used implements to cut the soil. The earliest was probably a tree branch or animal's antler used to break the ground's surface to produce a tilth for sowing seeds, but I am wrong to think this fetish for machinery I see in the eyes of some modern farmers is as old as the machine's invention. In 1939 the agricultural correspondent of the *East Anglian Daily Times*, responding to the government spending £1.25 million on a national tractor reserve, wrote that mechanization 'would encompass the ruin of the industry and cause desolation of large areas of land. It was essential to keep stock to supply humus to the soil.'[5] A Young Farmers debate the same year debated the motion 'Mechanized farming is more beneficial to the farmer today than horse labour' and was roundly defeated. Back then, farmers young and old seemed to think that good farming did not always involve the latest machinery. In Suffolk alone there were then well over 18,000 horses kept for agriculture, most of which had a working life of fifteen to twenty years, longer than most tractors, as the farmers had then pointed out.[6] A large number of these were Suffolk Punches: conker-coloured giants that grew up to 178cm tall and weighed over 1,000kg. Now there are fewer than five hundred pure-bred Suffolk Punches registered in the UK.[7] They are rarer, claim the society which protects them, than the Giant Panda.

We find something of this past in our crop fields one day. Passing one of the two old oaks that sit in the middle of the biggest field, I stop to look at its root ball. A hedgerow would once have joined these trees up. Soon, I hope, one will again. For a long time the trees have been ploughed within a metre of their trunk, the soil either side crumbling away so their roots are exposed, like blue veins on the back of old hands. I kneel onto the ground, and it is in this prayer position that I see something. I pull it from the earth and discover it is an old ploughshare. It is one hand long, curved at one end, and scabbed red and brown with rust. Behind me, in the still afternoon, the ghosts of the horses that pulled this plough begin to stir, as do the people who worked this land, the hedgerows and trees that once grew here, and all the creatures that lived within them.

The ploughshare is a biblical symbol of peace, and it feels important somehow to have found it.[8] Is this what we are doing, I wonder: making peace with land with which we have for so long been at war? That afternoon, when I get home, I put the ploughshare on a shelf in the sitting room, like a relic.

As spring breaks through the dawn creeps in earlier and earlier and with it the birdsong I have come to love. It starts in the darkness, growing louder every morning. I am beginning to learn the individual songs of the birds. When proper nature lovers spoke of the skill of recognizing one bird's call from another it seemed impossible that someone like me, who can no longer read music or speak another language, could ever do the same. But I download an app on my phone – a kind of Shazam for bird calls – and start to train my ear that way. There is no one in my life to teach me any more. Anyone who might once have known their songs is dead, so now technology will have to take their place.

I guess – record – pick a bird – play back. After a while I start to guess right: each time it feels miraculous. On one occasion I am walking with the children through a neighbour's maize strips next to a thick hedge when I hear a strange call I do not recognize. It is other-worldly, wheezy, like a 1990s synthesizer. It is coming from deep inside the bush, so that I cannot see the bird or work out exactly where it is. I shush the children and take my phone out to record its song. The three of us stare at the app's whirling red circle. A picture flashes up on the screen – *Bullfinch* – and I play the call back to check it is the same. It is. I peer deep inside the hedge and have to stop myself crying out, for inside I see a bird with an acid orange chest, Day-Glo bright against the brown branches, its black-and-grey striped wings folded over its back. I lift each boy up in turn up to see it before it flies further along the hedge's centre. We walk next to it for a while as it moves, listening to its squeaky-hinge sound and shouting 'There it is!' Afterwards, I will hear a bullfinch's call often and wonder how many times I have moved through a place without noticing it before. I have heard of plant blindness, when ignorance of a plant makes it difficult to even see it,[9] but I hadn't understood there could be bird deafness too.

The day I first hear the bullfinch is the spring equinox, falling two-thirds of the way through March. Now night and day are equal length, but soon the light will lengthen, banishing the dark. The need to mark this change has always been felt strongly by country people, right back to a pagan festival named after the Anglo-Saxon goddess Eostre, who was said to have represented fertility, birth and rebirth. Even now its Christian replacement, Easter, is still tied to it, always falling on the first Sunday after the full moon following the spring equinox.

On our way home the children and I see a hare

running along a neighbouring farmer's newly greened-up crop. Here in Suffolk they are neither common nor rare, although this is not the case in other parts of the country. Their numbers have fallen by 80 per cent in the last hundred years.[10] We have seen hares here before, but this time we are far closer to one than we ever have been. It stops when it senses us, freezing still. Time is suspended. I am struck by just how long its ears are and by the power of its legs and body, so much bigger than a rabbit's. I slowly bend down and lift Wilfred up to see it better and, in doing so, break the spell. The hare lopes off gently, unconcerned, its hind legs folding and unfolding like a lever behind it. The black stripe down its back and tail make it somehow more serious, less comic, than the bouncing white of a rabbit's. I tell the children that in Suffolk some people believe hares are witches who have transformed themselves so that they can play tricks or perform evil deeds without anyone knowing. As the dusk falls, their imaginations start to see the hare's true form in the long shadows of the oaks that line the field boundary. They are so sure that I start to see witches in the tree's dark reflections too. It is only afterwards that I realize the luck of seeing one that day. A hare, some think, was Eostre's consort.

Hugh sits across the table from us, looking at a map of the crop fields. The evening has come later than it did yesterday, and the day before. The seasons are shifting; growth has begun.

Hugh used to be an agronomist: an expert in crop

production. He's here as a favour to pass on his advice and experience as to what we might do with the new fields. Change is happening – both inside the cottage and out.

When the pesticide market boomed after the Second World War, so did the careers of those who made money selling them. In the UK, unlike other parts of the world, an agronomist's role is mostly, although not exclusively, to advise a farmer on the use of chemicals, and 88 per cent of arable farmers in the UK still use an agronomist to suggest what to spray and when.[11] Between 1990 and 2015 pesticide use in the UK fell by over 50 per cent, but their continued application over so many years has meant an increasing number of weeds have become resistant to the sprays designed to kill them, so farmers have started to increase their dosage. Between 2010 and 2018, pesticide use rose by 15 per cent.[12]

Although some agronomists are not now paid by the company whose products they recommend, a large part of their job is still dependent on farmers using pesticides. If farmers do not use pesticides, there is no need to pay someone to recommend which ones to buy. We have already converted the pasture fields to organic. Now we are thinking of converting the crop fields too. Doing so will put people like Hugh out of a job, so when Ben looks at me to start the conversation I am unsure how to go about it.

I am also suddenly very aware that I hold a competing set of powers and disadvantages. I am not the one who will drive the tractor, but I will still shape what happens in the fields. I am not the landowner, but I do get a say in how the land is farmed. I have come from the city, but also this life is in my blood. And, alongside all this, I am 'The Wife'. It is hard to shake the feeling that anything I say at this meeting must be backed up by someone else before it can be taken seriously.

Women farmers are, too often, still referred to as a 'farmer's wife' when male partners rarely become a 'farmer's husband'. Salesmen and union reps still ask if their husbands or fathers are in when they arrive at the farmhouse, and although she is just as able to answer questions about their farm, the microphone rarely finds its way to her first. Almost every woman I know who is married to a farmer but who is not a full-time farmer herself not only has an off-farm job but is also responsible for, or fully involved in, not just the farm paperwork and the majority of household duties but also lambing, corn carting, rolling, mowing, baling, tagging, milking, drenching, moving stock and a whole host of other duties that see her in wellingtons and waterproof trousers at five o'clock in the morning. There is no wifely job equivalent in any other industry.

This is still a world where, often, a certain kind of masculinity is prized. Still a world of 'boys' toys' and 'arable boys', where women care for stock and men ride the machines. Where a farmer might find herself bidding up to an overblown price for a flock of ewes because the other buyer cannot stand to be outbid by a woman. Where fathers advise daughters to complete an agricultural degree that sons have no need for because 'otherwise the men won't listen to you', or introduce their farming daughters as 'the son I never had'. Maybe it is this type of masculinity which makes so many farmers want to climb aside a bigger, shinier tractor than their neighbour, or which made artificial nitrogen so instantly popular, boosting a crop's growth so fast that a farmer would look over the hedge at the tall, stiff stalks in the next door field and think, *I'll have what he's having*.

This is just how it is. I am as aware of the constriction as I would be a tight pair of shoes. I blunder in, enthusiastic about all the things I have been reading about, listening to on podcasts and seeing on farms I have visited. I blurt out words and ideas. *Living mulches. Under-sowing. Minimum tillage. Companion cropping. Integrated pest management. Beetle banks. Mob-grazing.* I see Hugh's face grow red but cannot seem to stop myself talking. My palms prickle as they start to sweat. The room suddenly feels unbearably hot, like it has no air left in it. In the end it is a benign sounding word that is the trigger.

'I mean, I'm not suggesting we do this, or at least definitely not now, but the other day I was reading about agroforestry . . .'

It is too much for him. He throws his arms into the air and shoves his tall frame back in the chair, so that the legs scrape hard on the floor. The noise sounds like old car brakes. I do not finish the sentence. He is larger than me anyway, but now seems to take up twice the space. Afterwards, I decide that he didn't mean to start shouting, or really realize that he had, but that doesn't stop me flinching.

'AG-RO-FORESTRY? Well, Sarah, I mean . . . it's BOLLOCKS, it's all UTTER bollocks! I mean . . . how can it even WORK? How on earth do you get a bloody DRILL . . . let alone a bloody COMBINE . . . between a row of bloody TREES!'

I can feel his frustration flowing over the table in pulses. He eyeballs me, his finger jabbing.

'LOOK, I've been doing this a LONG time, a VERY long time. I've seen lots of things come and go, you know. And there's ONE thing for SURE I can tell you, which is this is all TOTAL rubbish.'

I flush red hot as a fat lump of humiliation rises in my throat. I look down, feeling like I cannot meet his gaze.

'You just REALLY shouldn't believe EVERY-THING you see on *social media*.' He pronounces 'social media' as though he has eaten something bitter. Then he pulls his chair back in towards the table with a bang and shuffles his papers. The matter is closed.

My eyes sting and I blink and reach for my tea, which I drink to try and slide the lump in my throat down. The tea is cold and it doesn't work. I say to myself, 'Don't cry, don't cry.' I feel like a total fool.

Agroforestry, I had read, was the practice of integrating trees within farming. Its technical name is silviculture: 'silvi' from the Latin for forest, and 'culture' meaning 'to grow'. It is a tinkling name, which sounds like spells and fairies, and I can see why those used to the thrusting titles of agrochemicals and modern wheat varieties might roll their eyes at it. Sometimes it can simply mean the planting of trees within hedges, or at the edge of land and rivers to provide a barrier and to use the tree's roots to help prevent soil and water from leaving the land. But there are other types too, depending on whether the trees are combined with livestock – silvopastoral – or with arable and horticultural crops – silvoarable. It is something that has been practised throughout the world for centuries but which fell away when monocultures and commodity production drove farming towards economies of scale.

I have learned that, when farmers are faced with something new, their first response is often, 'That's great but it won't work here. Our soil is too heavy / too light / too dry / too sandy / too wet / too thin / too coastal / too inland / too upland / too lowland / too different.' I have learned too that there is one advantage in trying to farm when you know very little about farming: every idea fits into the same category. You don't look for reasons why it wouldn't work because it challenges the way you've been farming until now. It isn't separated out under a subtitle of 'weird stuff'. And, this time, I do know that agroforestry might work here because I have found a farm less than twenty miles away that is already doing it.

The farm is called Wakelyns. Its experiments in agroforestry were started by Professor Martin Wolfe in 1994, after a forty-year scientific and academic career in plant pathology. Wolfe was a self-declared outsider. His distance from conventional farming, together with his academic and scientific understanding of plants, prompted him to look at how they might work together in a symbiotic farming system. The author of multiple published papers on the subject, Wolfe planted alternating rows of trees and crops north to south in 'alleys', which minimized the shade the trees cast on the crops.[13] He discovered that doing so had a strange effect on both. In one of the experiments – designed to find a solution to prevent the prevailing south-westerly wind from carrying airborne diseases such as potato blight onto the crops – Wolfe found that, while the first alley of potato crop was struck by blight, the second one, separated by a strip of coppiced hazel, showed far less. The further east the alley, the less blight appeared. The strips of hazel, he concluded, had acted as barriers to slow the spread of the disease.

In another experiment, the fruit and nut trees Wolfe planted between his cereal crops became so productive that he concluded it wasn't just the crops that were being protected: the trees, set apart from one another by twelve metres and flanked by hedges, were too. The key, he decided, was to intersperse plants, moving away from monocultures towards diversity.

Wolfe became convinced that there was some kind of communication taking place underground between the trees and the crops through mycorrhizal interactions – the tiny fungi which grow on a plant's root-hairs and become a go-between, a kind of underground broker, reaching further into the soil than the roots can and trading liquid carbon leaked out of the plant's roots for nutrients that the plant needs to grow. Even more curiously, Wolfe's frequent soil tests showed low phosphate and potassium levels, which would, in any other farming context, have required the farmer to add these to the soil. But the crops were doing just fine and producing increasingly good yields. Wolfe knew there was something else going on. He just didn't know how to prove it.

A number of studies have since supported Wolfe's theory that trees help rather than hinder crop productivity. They found that a tree's roots don't necessarily compete with the crops for water or nutrients because their root systems are deeper than the crop's and ensure a 30 per cent reduction in water evaporation from the soil. Further agroforestry studies have found higher levels of mycorrhizal fungi, larger earthworm populations and improved nutrient cycling. But, as well as this, by using the space above the crop, the trees have ended up making the famers more money, not less. A European research project carried out between 2001 and 2005 found that farm profitability increased by between 10 and 50 per cent with high-value trees such as walnut, but other farmers had found an additional income stream just from selling the timber.[14]

I don't say much for the rest of the meeting. I can't – physically – even if I wanted to because I fear my voice will crack and give my humiliation away. But sometime later, I listen to a *Farmers Weekly* podcast. The person being interviewed claims that agroforestry is becoming one in a 'suite of options' for the future of farming. The next edition publishes an article entitled: 'DEFRA Paves Way for More Agroforestry Projects in England'. It ends with the declaration: 'The environmental benefits of agroforestry are clear and well-established, and will be the primary driver for change in coming years.' *Huh*, I think.

A while later I visit a Cambridge tenant farmer, Stephen Briggs. He farms 250 acres of cereals and vegetables on fenland peat soil. In 2009 he and his wife planted 4,500 fruiting and juicing apple trees in between his crops in an effort to try and stop his fine Fen soil blowing away.

Stephen Briggs looks and sounds like any other farmer. He would look at home standing in the John Deere tent at an agricultural show or at the bar of a country pub, but this is not a world he was born into. He and his wife both trained as soil scientists, but no one would give them a farm tenancy. After applying and being turned down several times, they eventually put together enough money to buy a small plot of land with which to prove their farming credentials. It worked. A council offered them a fifteen-year lease. The farm had conventionally grown wheat, oil seed rape, sugar beet and potatoes in sequence and had a bad weed problem. But Stephen and his wife had no money to establish an arable farm from scratch, with huge sums needed for pesticides, artificial fertilizer and machinery. Their only real option was to farm organically. For a long time, the biggest piece of kit they owned was a ride-on mower.

Stephen's fruit trees are planted in alleys, twenty-four metres apart. He has spaced them out like this so that his modern farm machinery can still comfortably fit between the trees. The length of his tenancy means he needs the trees to earn their keep quickly, so he has chosen to grow fruit trees alongside his arable crops to provide him with an annual and a perennial income. In between the trees he has planted a mix of clovers, vetch and wildflowers to suppress weeds and attract insects to pollinate the trees and eat pests on the crops. Pollinator insect numbers are now 400 per cent higher than when he started measuring them.

The trees take up only 8 per cent of the land; 92 per cent of it is still given over to arable farming. In a normal weather year, he still makes more or less the same amount of income as his organic neighbours. Recently, though, there have been no normal weather years. So now, when the Fen wind blows sharp and bitter across that treeless and hedgeless landscape, or when the rain falls for days, or when it doesn't fall at all, then he really sees what they offer him: resilience.

But it is what he says about the connection between tree and crop that is the most extraordinary. Not just that his yields have gone up 25 per cent in ten years, but that the crops nearest the trees grow noticeably higher than in the centre of the alley. Often, when he is harvesting, he has to tilt the combine header next to the tree lines because the crops that grow alongside them are so much taller than in the middle of the alley. Something strange is happening underground.

To Stephen, there is nothing magical about this. His job, as he sees it, is to harvest sunlight, mix it with air and water and build carbon, which he then sells as food. Agroforestry allows him to farm a bigger area below ground by using the tree's roots, and a bigger area above ground using the tree's canopy. He calls it 3D farming. But he also knows where he, the farmer, stands in this. 'Nature is far smarter than us,' he tells me. 'It has worked out all the niches a long time ago . . . we simple humans are just playing catch-up.'

When I visit his farm I watch Stephen walk a few metres into one of his fields and dig into the earth with a spade. He brings up a clod of soil and takes it back to the group of visitors I am standing with at the field's edge. We crowd around it, looking at the dark soil filled with roots and worms. Clearly visible along the fat roots, which reach out like fingers, is a white fuzz, which looks a little like miniature icicles running along a frosted window ledge. The white fuzz is mycorrhizal fungi.

Amongst the group are two agricultural students in their third and final year of university. They ask to look a little closer. They've never seen mycorrhizal fungi before. None of the farms they've worked on have had enough for them to see it with the naked eye.

Stephen's language is one of economics and practicality rather than ideology. And the truth is that, even though Wolfe and those like him did much of the research and bore all of the ridicule, it will be farmers in Schöffel gilets who will convince other farmers to change. And they are. One, John Pawsey, a neighbouring Suffolk farmer, puts it this way: 'Agroforestry is a bit like getting a tattoo,' he says. 'First you put it where no one can see. Then before you know it, you're covered in it.'

Much later, Stephen will become one of three farmers nominated for *Farmers Weekly* 'Arable Farmer of the

Year'. 'In conclusion,' the article about him says, 'while many said his system wouldn't work, Stephen has proved them wrong.' I think of Hugh and wonder whether maybe we might be able to do the same.

The summer is hot, with long, endless days. I fail to realize that it is to be our last normal summer for years in ways I could not have ever guessed.

It was once said that harvest began on 1 August with the Celtic tradition of 'Lammas' – 'loaf mass' – when a loaf of bread was made from the new wheat crop and given to the local church. I learn about another old tradition named 'calling the mare'. When a farmer finished reaping his harvest, the last sheaf was woven into the rough shape of a horse. To show he was faster than everyone else, the farmer would throw the corn mare into the field of a neighbour who had not yet finished, crying 'Mare! Mare' before running off. This second farmer then had to work fast to make sure he finished before another farmer did, then throw the mare to them. The final farmer in the parish to finish had to keep the mare all year, displaying it so all knew he was the slowest. The same idea is now played out in village pub bars all over the country with the question 'Got yours in yet?'

Harvest once had to be completed by Michaelmas, when the winter curfew would begin. Each night the church bell would toll, one strike for each of the days of the month that had passed, rung every night except Sunday until Lady Day on 25 March. Now the children and I ignore the tradition that blackberries must not be eaten after old Michaelmas Day, when Lucifer was said to have been thrown out of heaven, falling from the skies into a blackberry bush, which he cursed, breathed fire at, stamped and pissed on to make the fruit unfit for eating. We take a plastic tub and collect them anyway, but when I tear the skin of my thumb on a thorn, sucking it to stop the blood, I wonder if I should have paid more attention to the folklore.

The autumn light is luminous, as though someone is filtering the sun through muslin. Farmer Martin's cows are now allowed to graze with ours in the meadow fields. I find myself standing and watching their backs glow under the sun, listening to the sound of their grazing as though it were a meditation, which, I decide, it sort of is. I watch them strip the lower leaves of the old white willow tree, whose bottom branches sweep down into the field. We nearly cut these branches off in case they broke the fence but then I read that as it ages a tree will put down a branch like an old man might use a stick, to steady itself, and we decided not to. Now I see how the cows use the leaves for food and the branches as a scratching post. They reach high into the tree when they've eaten the leaves on the lower branches, curling their tongue around them to pull them off. A farmer tells me that willow has a form of natural aspirin in it, and that her cows too like to eat it before bed.

I try to organize a harvest tea like the ones I used to have when I was a child. I want my children to have their own memories of the feeling of stubble straw poking through a rug and the specific taste of sweet tea drunk from a plastic thermos cup. But harvest now is too efficient: we are disorganized and miss the chance and the crops are cut without us. I feel the absence of marking this change of seasons in a way I do not remember noticing before. I am aware of time shifting: the end of one season and the briefest of pauses before the next begins. The harvest moon rises over the meadow field, so huge and low and orange it looks unreal, like an airborne space hopper. It falls every September, its light so bright that once farmers working in the fields would have used it to finish harvest after night fell. Now mighty combine headlamps shine out beams several metres long and their drivers work until the early hours of the morning. Even so, when the harvest moon comes, I take the children outside in their pyjamas and boots and we all stand in its light and howl at it.

There is another change coming. The London house is nearly ready. Ben now has work that sees him travel regularly to London, and Suffolk is too far for it to be a daily commute. We agree we must move back and I make myself excited at the idea by thinking of friends, parties, restaurants, galleries, theatres. We want to keep on managing the farm and see through the changes we have started, so we will come up every weekend, half-term and holidays, staying with Ben's parents in his childhood home across the pasture fields. I tell myself how very lucky I am to have had this chance at all.

I advertise the hens on Gumtree – *CHICKENS: FREE TO A GOOD HOME* – and a lady in a blue van comes to collect them. She is warm and friendly and

I am glad. Afterwards, she sends me pictures of them pecking around her stable yard, as though she's a babysitter placating a worried parent.

As I pack up I think how, although this place may not be part of my childhood, these years have made it part of my children's. Their experiences here have taught them lessons they will not forget. No matter where they end up living, it will always be a part of them, just as where I grew up is a part of me. This makes our time here feel very important, as though I have a duty to remember it for them. In the book *Kith*, Jay Griffiths writes about how our childhood landscapes can define us:

> I was interested in how children belong, needing their kith, their local acre, as they need their kin. An entire history of childhood is in that one word 'kith', which is now used as if it means only extended family, whereas in the phrase 'kith and kin', 'kith' originally meant country, home, one's land. Childhood has not only lost its country but the word for it too: a country called childhood.[15]

I think about the idea of 'a local acre'. Someone once told me that people's accents were once so specific that people could tell a visitor from a neighbouring village just by the inflection of his vowels. We have lost 'kith' from our vocabulary, but we have also lost differences in the way we said it. In a BBC documentary about the accents of prisoners of war recorded between 1915 and 1918, dialect coach Joan Washington said she had come to believe there was a connection between the way people speak and the landscape that surrounded them.[16] She loved travelling to farming communities because they tended to hold on to their accents and dialects far longer than others did. An accent, Washington believed, enables us to get into the sensibility of people. It can define who they are. But what about when we lose our accent? Do we lose those sensibilities also? I wonder whether learning the landscape has managed to uncover some of my true sensibility and helped bring me closer to my kith and kin, for I feel the loss of leaving here in a way I had not expected to. The farm and this land has inched its way under my skin. It has connected me to the natural world in a way I have not been since I was a child. I am no longer outside nature looking at it; I am inside and part of it. Invisible threads have joined themselves from me to the world around me. I understand two things: that the threads are all connected, and that there are hundreds of thousands of them of which I am still unaware.

I wonder whether, when I return to the city, those threads will be broken. Living alongside 9 million other people is like a drug: the greater your exposure, the greater your tolerance. It is impossible to absorb it all and so you dial down your response. You build an armour, loosening and tightening it when you need to, although the trick of doing so can be a difficult one. I wonder whether I will now start looking down instead of up; begin blocking out noise so I only hear the loudest sounds. I wonder whether it is possible to continue to see the details of connection when there is just so much to look at. I am not sure, now that I have begun to see it, whether I am really ready to stop.

I know, though, that I am changed. As I pack up our clothes I realize we are not moving back to an old life with our old things but moving to a new place, as new people. I can see now how all of this was possible only because of the destruction that came before it. The last time I packed and unpacked it was in the midst of joblessness, uncertainty and confusion about our future and our path. I could not have known then where any of this would take me, nor how glad I would be for its lessons.

The same lessons were learned by two farmers I have come to know. Like me, Rebecca and Stuart found that something they had thought was an end proved to be a beginning. Their journey, like mine, taught them to look at their world completely differently. I learned a lot about farming from them but I also learned that sometimes it can be the things we think might destroy us which actually end up doing the opposite. We both just needed to learn a new way of seeing.

*Everyone has to look in the mirror and say, this is me,*
*I'm proud of what I do.*

Stuart, farmer

# 8. Rebecca and Stuart

They almost never leave the farm, but it's not like they can refuse to be at the wedding. It's family. They have to go. And it's not like they can ignore the request to switch off their mobile phones either. They are pig farmers, not heart surgeons. Producing pork is not a reason to interrupt someone's nuptials with the ping of a text. So they never get the message that the alarm is ringing.

The first person who hears it is the farmhand. He arrives that day in the bright blue of a perfect September morning. The light falls with the strength of late summer and everywhere is bathed in it, even the pig sheds. This is Norfolk light: the light of paintings and poetry. It is the kind of light that makes artists set up easels before horizons of uplit clouds and look up to the sky in reverence, half-expecting to see God there looking back at them. It is this light that now falls on the frowning face of the farmhand as walks across the yard, confused as to why he can hear the sound of ringing coming from one of the three pig sheds. The sheds were built by Stuart's parents in the year that he left school, after the battery hens they had been rearing began to fall out of favour. Each shed houses nearly a thousand full-grown pigs, although in this age of bigger-betterness, there is nothing especially unusual about their size.

The farm had begun to breed pigs long ago. Lots of farms did around here as the free-draining Norfolk soils meant it was easier to keep the animals outside without having them churn up the land. But as the years passed, more and more farmers began to bring the pigs inside as it was easier to maximize the speed and efficiency with which they gained weight this way. Housing thousands of pigs took up far less space than if they had been out on the 500 acres that made up the rest of the family farm. This could then be used to harvest crops instead.

Pig business, unlike most other areas of farming, has never been subsidized. It has, therefore, always been a case of boom or bust. Its prices, supply and demand rise or fall on the waves of the international market. If disease wipes out half of all Chinese pigs then you are back in business making piles of sausages to keep the masses happy. But if too many farmers think the same, over-supply soon means pork prices come crashing down again. Stuart and Rebecca, like other pig farmers, were used to the back and forth of it in a world where food was now a commodity.

The money to buy the sheds had partly come from selling off the old farmhouse that the fields once fanned around. Now Rebecca and Stuart live in one of two new modest red-brick houses, one for them and the other for his parents. The rest of the money had been borrowed, like it is by most farmers honest enough to admit the cost of progress. But once the bank owns the asset, the asset must be sweated. Bank managers talk about delivering 'economies of scale' and the ends begin to justify the means. And so it isn't until the money is finally paid off, and all that is left for the work is a modest annual income and an out-of-date building, that such an enterprise can begin to look less like an opportunity, and more like a trap.

The farmhand assumes the ringing must be a fault with the system. When he gets to the shed door the noise is so loud that he opens it as fast as he is able. In as much as he had ever thought about it, he had supposed that carbon dioxide would be odourless. But it's not. When concentrated, it smells a little like acid: putrid and fierce and dangerous. It is this smell that warns him but still, he opens the door and goes inside the shed and it is only then, when he sees the dead pigs, that he fully understands what has happened. Others might have fled. Instead he holds his breath and runs as fast as he can down its full length, flinging open the two internal doors within it and the one at the other end before collapsing into the sunlit yard. Afterwards they tell him that there was so much gas trapped inside the shed from all the dead pigs that, had he stayed inside it any longer, he would probably have been killed by it too.

The ventilation system failure has filled the barn with the carbon dioxide of three hundred and fifty full grown pigs who fell asleep and never woke up. Half the shed is dead. Stuart and Rebecca's insurance company value the loss of the animals at around £50,000. They won't pay out because the alarm had worked. It was just that no one had heard it.

Every farm has death on it. After all: where you have

livestock, you have deadstock. Stuart and Rebecca both know that. They are not sentimental or whimsical about what being a farmer means. Both have seen death and disease up close and personal. They know it to be a part of life: both the animals' and their own. But everyone has their limits. In hindsight, it's hard to choose exactly what it was that broke them. Layer by layer, the weight of it all built up until, in the end, it cracked them open.

The first cracks had begun to appear ten years earlier. At the start of the millennium, disease after disease had washed through the pig sheds. Porcine dermatitis and nephropathy syndrome. *CRACK*. Post-weaning multi-systemic wasting syndrome. *CRACK*. Swine flu. *CRACK*. Foot and mouth. *CRACK*. At its worst, their mortality rate was at 25 per cent which is, if you stop to think about it, a lot of dead pigs.[1] *CRACK*.

When the ventilation system fails in 2010, the farm workers, worried for their own jobs, are sure that the family must finally quit. They presume that after all the death the decade had delivered up, it will be impossible for the family to keep going. But the truth is, they cannot quit. Their minds have to refocus on the looming shadow of debt and what must be done to keep up with repayments. The monster must be fed. So they simply begin again.

They keep going for five more years before the next *CRACK* comes. In the spring of 2015 another disease finds its way into the sheds. It is known most commonly, like so many of them are, by its three-letter abbreviation. APP. *Actinobacillus pleuropneumonia*. Over the next few months they lose the entire herd. Every single animal in every single shed. Over two thousand pigs, dead and wasted.

Once your life has been cracked apart it's hard to know what to do next. And so Stuart and Rebecca and their two small children do something that livestock farming had, up until then, made impossible. They go on holiday.

They make their way to the mountains of Scotland, where the summer light can sometimes look a little like it does at home. Here they clear their heads and hearts, and try to work out what to do.

Maybe if Rebecca hadn't seen the cows, things would have turned out differently. Maybe they would have come home and just tried again. Or maybe they would have been able to deny that part of themselves that made them farmers: sold up their share of the farm, paid off their debts, and got out. Had they done so, their lives would now be very different: better, but also worse.

Instead, they stay in Scotland on a friend's farm. Their friends have Jersey cows the colour of honey, with wide swinging hips and eyes that look like they've been ringed with kohl and given Maybelline eyelashes. They are the bovine equivalent of a seventies supermodel. Rebecca sees the cows and falls immediately in love. Being based on love, the decision to buy a couple of the pregnant cows and arrange for them to travel hundreds of miles to the east of England is taken on impulse, rather than careful planning. It is utterly mad. It is the thing that will eventually keep them sane.

*

They had left their farm as pig farmers with no pigs. Now they return as dairy farmers with no dairy. They have three empty sheds, five hundred acres of land and an overwhelming need to do it all differently.

The pregnant Jersey cows arrive that autumn, but it is only when the calves are born that they actually talk about what will happen next.

'Well,' says Stuart, 'you take the calf away at a day old, maybe two, once it's had enough colostrum.'

'I'm not doing that.'

Rebecca is not an overly sentimental woman. She is not prone to anthropomorphism. Afterwards she will wonder if it was the cancer which drove her out of her job as a land agent that made her look at life differently. She wants to give the disease no credit – no credit at all – but I cannot help but wonder whether her illness, and her recovery from it, changed the way she saw things.

'I'm not taking the calf away. Not doing it.'

'But then you can't milk twice a day. You lose half your sellable milk if you don't.'

'Well, what do you feed the calf on then?'

'Milk powder?'

Rebecca laughs. 'And how much does that cost? Not much less than what you'll get for selling the milk, surely? Dairy farmers are always bitching about how little they get paid for milk, at the same time as bitching about how much milk powder costs. I mean, why don't you just let it drink its mother's milk for free?'

Stuart thinks. He knows from two decades of pig breeding that you wanted piglets to get as much of their mother's milk as they could. It stopped them getting sick. It helped them to fatten up. It was so important that, now commercial sows had more piglets than teats to feed them, some farmers found themselves running between pigs trying to cross-foster piglets. Few farmers ever went to the expense of setting up a powder feeding system.

'It's just . . . it's just what you do. That's what everyone does.'

She looks at him. 'Don't care.'

'Well . . . what else can we do?'

'Don't know. We'll just have to figure out another way.'

They do figure out another way. They find out that a tiny number of dairies leave the calf with its mother. The calf is weaned at around six to ten months, when its nutritional needs are met by grazing. If the cow is losing condition too quickly because her energy is going into feeding the calf, then the calf is weaned earlier. Sometimes the cow will wean the calf herself, moving away when it tries to suck, pushing it back, fed up with the butts and sucks to her udder from a calf big enough to survive on grass. Other times, if the cow is doing well, then the calf is just left to wean itself. There are no hard-and-fast rules; no timetable or schedule. They just have to watch and respond. But this way of dairying means they can only milk the cow once a day and they don't have nearly enough cows to make this pay on a commercial dairy contract. So they decide to sell their own milk – unpasteurized and unhomogenized – at the farm gate. To do this they need a shop, and so they open one.

Rebecca leans against the cow's butter-coloured flank and teaches herself how to milk with second-hand equipment and YouTube videos. They set the parlour up in one of the empty sheds. It buzzes with strip lighting, the ghosts of the dead pigs and their old life.

But none of this – the dairy herd, the new farm shop, the arable acres – will satisfy the demands from the bank, which arrive with clinical regularity. The cavernous empty pig sheds must still be paid for. Their only real option is to go back to keeping pigs until the rest can pay its way. This time, though, instead of breeding and rearing their own, they raise weaners for a pig company: four thousand of them at a time.

The following spring, there is another crack. On the day that it appears, the sun shines strong and warm on Stuart's face as he walks across the yard. He doesn't realize that the ventilation system in the pig shed has failed for a second time because the alarm has failed too. There is nothing to stop him opening the shed that morning and finding sixty piglets dead. Without speaking, he turns away from the bodies. As he walks across the yard in the spring light he pulls his phone from his pocket. When he makes the decision, his mind feels clear. When he says the words 'no more' to the fieldsman from the pig company, the words feel like a yoke has just been lifted from him.

The fieldsman arrives just over two hours later. He must, they think later, have driven like a maniac to get to the farm that quickly. But the fieldsman wasn't thinking about whether or not he was driving like a maniac. He wasn't even really thinking about the dead pigs, nor whether Stuart would refuse to keep the ones that had survived. Instead, he was thinking about the gun cabinet in Stuart and Rebecca's house. He was thinking about whether he could get there in time to stop Stuart unlocking it, taking a gun, and then his own life. He was thinking about this because it had happened before. It happened all the time.

Death. Destruction. More death.

They aren't doing anything especially different from anyone else in the business. It is just the accepted cost of raising animals like this. But losing animals looks different in person to how it looks on a balance sheet. It looks like sores and calluses and stumbling. It looks like rasping breaths and not wanting to stand. Mentally, it looks different too. It looks like: *I should have noticed something; I should have been able to save it; I failed them.* Almost every farmer holds two simultaneous thoughts: animal death is part of farming; animal death is hard. Raising an animal for a purpose makes this feeling tolerable. If the animal's death has no purpose, the waste of it can be difficult to bear. It is hard when it is one or two or four, but it is also part of farming. Farmers harden themselves against it. They do it for us, so that we don't have to. But when the death is piles of bodies, bloated and stiff, it is not just hard. It is trauma. There is only so much of it than anyone can take before it seeps through the layers of toughness and into someone's bones. Later, Stuart will joke to me that all that death has given him post-traumatic stress disorder, like a soldier returning from war. Listening to him, though, I don't think it is a joke. And I also don't think he's the only one. It's just that, for some others, no one drove fast enough.

It all starts when they must work out how best to feed the cows, never having kept cows before. To begin with, they keep running out of grass as the seasons they rely on for rain and sun to help it grow bring floods and drought instead, and so they start reading books about how best to grow it. Books about grass are, of course, also books about soil, and so they begin to read about that too. Because of this, both of them trip and fall down a rabbit hole of learning that neither had foreseen. Maybe if they had taken this journey two years, five years, ten years earlier, both would have found it impossible. The ideas and concepts and challenges they were being presented with would have been too counter-cultural. It would have been too difficult to unlearn what they had been taught. It would have been too hard to accept all the things they didn't know that they didn't know, something that Stuart will later call *ecological illiteracy*.

As they are figuring all this out, the final crack appears and blows them open. It is the summer of 2019 and Stuart is standing at the edge of one of his fields. This year will see the UK's summer and winter temperature records broken. The field that Stuart is standing at the edge of has seen just a quarter of its usual monthly rainfall. The Met Office will use words like 'exceptional' and 'unusual' to describe a year of extreme heat and rain and cold and wind, but when every year seems to be a record breaker in one way or another, it's unclear when the forecasters should just start to call it 'normal'.

Stuart is watching a tractor pulling a plough, which is turning the earth, readying it for crops to be sown. Seagulls trail behind it, swooping and landing and fighting, their whiteness stark against the flat landscape's vast brown field. But Stuart is not looking at the birds. He is not looking at the part-time tractor driver sitting in the cab. Instead, Stuart is looking at a wall of dust stretching across his flat field, streaming either side of the plough and above it too, blocking out the tractor's rear window and reaching halfway across the field. It is a wall of dirt; a soil tsunami.

The tractor turns at the headland and, as it does so, the wind changes direction. Stuart sees the dirt cloud come towards him moments before he is engulfed by it. It catches in his hair, eyelashes, throat. He turns away, coughing, and walks across the field margin, through the yard and back to their house. He goes to the kitchen sink and turns his bare forearms under the tap. The water flows brown. He must have done this a hundred times before without noticing. He must have seen it every year of his life. He is the son of a farmer who is the son of a farmer. They have farmed this land for seventy years. But this time, it's different. This time he really sees what he is looking at. *That's our soil*, he thinks, watching it run down the drain in a miniature whirlpool, gurgling around the stainless steel. *That's our soil.*

The rest of the day is both uneventful and momentous. Stuart goes to find Rebecca and, as he walks towards her, she can tell something has changed, for the two have learned to read each other's faces. They have had to. Sometimes, over the last few years, words haven't been enough.

'Come and look at this,' he says, and takes her to the edge of the field. They stand and watch their soil blow away.

'That's the last time that's going to happen,' says Stuart. And he is right.

When Stuart and Rebecca first begin to hear about 'mob-grazing' they wonder if it might be something they should try. It seems to make sense. And after all, they've broken so many farming rules they don't see why they shouldn't just keep going.

They learn that this method of grazing animals is rooted in the history of the Great Plains of North America, where herds of wild buffalo once moved across grassland, creating some of the most fertile land masses on earth. They would do so as a mob – bunched up together in a herd, or pack – to better protect the weak and young amongst them from predators such as wolves. They grazed an area then moved on, all keeping together, staying away from where they had just been. When the colonists arrived, the buffalo were hunted to near extinction. The soil was farmed of all its fertility until, in the 1930s, it simply blew away and turned the land into a dust bowl.

This method of grazing was not confined to the Great Plains. A century ago, many of the fields in the

UK would have been small and hedged. The animals within them were moved regularly onto fresh pastures because the old farmers knew to *never leave sheep in the same field long enough to hear the church bell ring twice.* They had learned what indigenous peoples the world over had understood for generations: grazing animals don't wreck the soil. They build it.

What they didn't know was how.

For many years our understanding of soil was based on the ideas of a nineteenth-century German chemist, Justus von Liebig (1803–73).[2] Soil was understood to be disintegrated rock modified by weather. Soils were static. They had a kind of internal 'balance sheet' so that when nutrients were removed by harvesting crops, they had to be replaced. Most of the early work on soils was done by geologists because they were the people skilled in the relevant scientific methods and in possession of the right equipment. And so it was geology, rather than biology, that decided how soil was to be seen.

But nearly two hundred years ago, at around the same time that Liebig's 'balance-sheet' view of soil was being formed, the biologist and naturalist Charles Darwin concluded something radically different. He suspected that soils were actively changed by a complex and symbiotic relationship with animals or plants. Bioturbators – for example, earthworms – changed the soil, altering the availability of resources within it to other species. Because of this they, and other bioturbators, have been called ecosystem engineers, with the ability to create biodiversity within an ecosystem. Charles Darwin wrote his final book on the subject – *The Formation of Vegetable Mould Through the Action of Worms*. Although few today know of or have read it, at the time it was published it outsold the book for which he is now more famous, *On the Origin of Species*.

It wasn't until the 1980s, around a hundred years after Darwin died, that soil scientists began to question the long-established view of soil formation and the role that biota played within it. They began to accept that maybe it was the biologist, rather than the chemist, who had been right all along. Even now the majority of microbes that are present in soil have yet to be discovered, which is less surprising when you learn that a gram of soil can have from a million to over a billion of them living in it.[3] These microbes have a profound impact on the soil, influencing not just its pH but whether the nutrients within it are available to plants. What most soil scientists seem sure of is not only that these microorganisms play a huge role in a plant's life and relationship with the soil, but also that we are only just at the very brink of really understanding how they work.

Stuart and Rebecca begin to learn how the soil and its biology are changed by grazing. They learn that when a plant is eaten it starts a boost of growth – partly stimulated by enzymes within an animal's saliva – in an effort to repair itself. To begin with, the plant puts all its energy into pushing out new leaves, not pushing down new roots. If an animal goes back to graze these new sweet shoots, the cycle repeats itself and, over time, the shallow-rooted plant weakens. But if the plant is allowed to keep growing without being grazed it will grow deeper roots to feed its larger leaves. The bigger the plant's roots, the more carbon it can store within them. Some of this – some think up to 40 per cent of all the plant's photosynthetically created carbon – leaks out of the roots into the soil. This liquid carbon is then traded by microbes for the nutrients that the plant needs.[4] The world created around this one plant's roots impacts everything else in the soil.

Today there are few small fields left and even fewer wild predators to ensure animals move regularly onto fresh grazing so farmers are finding a modern way to recreate what would once have happened naturally. They take a field and divide it into sections with portable electric wire. The animals graze in the first section. Because they don't have the whole field, they get competitive. They don't just eat the sweet grass but the rough stuff too, trampling their muck into the ground as they go. When the animals have eaten a third, trampled a third and left a third, they are moved into the second section and kept off the area they've already grazed.

Stuart and Rebecca discover that every mob-grazing farmer varies how often they move their animals depending on the weather and the season. Most move them after a day or two. Some move them every day, or twice a day, or even more often than that because they have become addicted to what it does to the grass, which grows and grows, high enough to brush against the shoulders of the cows and bury the sheep as they wade through it. According to the UK's Agriculture and Horticulture Development

Board, grazing animals this way can increase the amount of grass in a field by between 28 and 78 per cent.[5] But farmers don't do it just because there is more grass. This is deeper rooting grass. It will survive droughts and floods when other fields do not and will enable the field to drain when it rains, increasing biology in the soil all the while.

The old farmers did not use terms like 'root exudation' or 'rhizosphere'[6] or 'mycorrhizal fungi', but they did know that if animals moved regularly onto fresh pasture then the grass would grow better and the animals would look well. Or maybe they did use those words, but they couldn't start a blog or a podcast or a YouTube video and no one asked them to write it down, so they just passed on the knowledge in the old-fashioned way. They told their kids.

When the chemical revolution came along promising miracles, and delivering them, the kids stopped listening to advice about the old ways. They stopped listening long enough for them to nearly be forgotten. So Stuart and Rebecca find themselves having to learn this from books and videos and podcasts. Rebecca discovers a WhatsApp group of farmers who are all mob-grazing their animals. Over the course of a year its membership grows from four to over 140. They share advice on practical husbandry, getting water to the animals or different species of plants to grow. They share how their cattle, given the chance, will graze not only on grasses but also on hedgerows and trees. They learn that trees were once pollarded in summer – the blowsy branches cut and stored in a barn until winter – the 'tree hay' given to hungry cows to strip bare, supplying them with additional minerals and nutrients. They post videos of cows eating thistles and docks and all kinds of weeds Rebecca and Stuart had always understood they would refuse. They learn that the more types of plants they have, the better the animals seem to do. One farmer sends a video of a herd he has kept out over winter for the first time. They are heavily pregnant but in such good condition that they skip and buck when let into a new block of grass, making him sure that none of them will need help with calving, for it is mostly the fatter animals that struggle and need their calves to be pulled out with ropes or winches. Every now and then someone will write, 'This group challenges everything I had understood and been taught about farming'. *Yes*, think Rebecca and Stuart, *it does.*

The group share another accidental consequence of this kind of grazing. Seeing their cattle and sheep and other animals up close each day – even for the five minutes it takes to move them from one section to another – has changed their relationship with them. They spot problems more easily. The animals become used to regular human contact and easier to handle. One member sends a video of a Canadian rancher, Neil Dennis, able to lead 1,000 cattle huge distances with only his wife to help him close the gates.

What Stuart and Rebecca once thought was normal – a big, fenced-off green field of ryegrass dotted with cows in every direction – now starts to seem bizarre, self-defeating, with nothing for the animals to self-medicate or balance mineral or nutrient deficiencies by themselves. They realize that this type of farming is not just changing the way they think. It is changing the way they see.

In 2020, when the new year comes with promises of beginnings, Rebecca discovers a farming course she thinks they should go on. The course is entitled 'holistic farming', which makes Stuart think of hippies and crystals and instinctively he wants to mock it. They go back and forth on the cost but then they find out that 'holistic' derives from the Greek word, *holos*, meaning 'whole', which comes from the root word, *sol*, meaning 'whole', 'well kept'. A well-kept farm? A farm in which each part of the system works together as a whole, linked, joined up? Well, he supposes he could get on board with that.

What Stuart and Rebecca learn on the course changes them. It changes the way they think. Some call it a paradigm shift, like something from *The Matrix*. The amount of learning and information is both terrifying and thrilling. This is a world that tests soil to see not what needs to be added to it, but whether or not it is biologically alive. When faced with a disease or pest it asks not 'What will kill it?' but 'What will eat it?' It encourages them to re-find their own instinct. Each decision they make must be turned up to the light and examined to make sure that it is the best one economically, but also environmentally and socially too. It must work for every part of the farm and, as they are part of the farm, it must feel as right for them as it does for the cows or trees or birds or bank balance. It asks them to look up, to look down, to watch and listen and, in doing so, learn. It asks them, therefore, to connect with the land they are working and really see it again, but this time as though they are part of it.

The growing popularity of the farm shop means that they have had to start to sell more than just milk. As well as the dairy herd they now have goats and sheep, Herdwick ones, so hardy and old that their name can be traced back to the Old Norse *herdvyck*, meaning 'sheep pasture'. They have pigs too, except these ones are hairy and robust rather than fast growing and weak. They have built up a beef herd, a motley crew of native breeds that grow fat just from pasture and need little looking after.

They discover they can graze the beef cows on much more than grass. They graze them on the winter stubble after harvest, letting them eat leftover crops grown up from seeds dropped by the previous harvest. They eat blackgrass, the weed that plagues other arable farmers but which is, as Rebecca says, after all just grass. Each year Stuart and Rebecca walk across a field and realize how much more there is growing in it than the year before. Where the spring and summer had once turned up dirt or docks or thistles, now there is grass. Where parts of the field were once boggy or poached, now there is grass.

They stumble across accidental consequences all the time, both good and bad. They mow a path for visitors through some planted meadow and realize how the cutting enables the clovers, which had been struggling, to thrive, bloom and grow. They sow wildflowers to encourage wasps in order to pick off flies from the cows in the summer, then watch helplessly as the same wasps kill two hives of bees. A field right next to the road, which has been permanently cropped and suffered from bad drainage so that there was always a puddle in its middle, is sown with a mixture of plants that shimmer with the purple-blue of chicory. For the first time in years the puddle disappears and Stuart is no longer asked if he has 'the farm with the field with the pond'. They grow a pumpkin patch and parents bring children to pick them, then ask Rebecca if the vegetables are safe to eat because they have never seen pumpkins 'growing in the wild before'.

They begin to realize that part of what they must do as farmers is not just to sell milk and food, but to connect people to what they eat and to the land. They start inviting the public to watch them milking the cows. The first person to do so is a man in his early seventies who stays and watches for over an hour. He asks Rebecca question after question and she tries to answer while silently praying that the cow she is milking will not perform her usual trick of shitting everywhere. Her prayers don't work, but the man will come back anyway.

They are frequently stretched to breaking point. In spring, when goats are kidding and sheep are lambing and cows are calving and everything is coming all at once, it can feel impossibly hard. In the midst of the chaos their first Jersey cow – the one who started it all – dies of milk fever. Rebecca finds it almost impossible to forgive herself.

It is hard. It is hard not to feel overwhelmed with new information. It is hard to farm all day and learn all night when there are small children who need to be fed and watered and loved. It is hard to keep animals again because, when there are problems, Stuart's heart races and his head fills with the ghosts of dead pigs and he has to walk away and let Rebecca deal with it alone. It is hard doing this with your land when so many around you are not. And although there is some government money available, it is also hard when the only reward to improving your soil fertility, your water drainage, your trees and hedgerows and the numbers of insects and birds you see each day is the ability to tell yourself you're *doing the right thing*. One day, though, Stuart walks some of the fields with his parents. The farm looks different now. It is alive, no matter what the season. His mother turns to him and tells him something he never knew: she and his father called him Stuart after 'steward of the land'. So now Stuart the Steward has begun to fulfil his prophecy.

It is Rebecca who takes the call from the man who books the final table in the new restaurant, next to the farm shop. The booking is for Mother's Day: Sunday 22 March 2020. She texts Stuart.

*Fully booked. Whoo-hoo!*

They have been advertising it for months. It has been a slow process, building up local custom, persuading people up the path to the farm. Now a new car park and building houses a new kitchen and series of tables all set and ready for diners. Behind it is the refrigeration unit and processing area, where Stuart has taught himself butchery from YouTube.

The BBC announcement comes up on Rebecca's phone. Its jingle is a series of beats, like a heart racing.

UK PM Boris Johnson announces closure
of pub, bars and restaurants.

She is surprised when she doesn't cry, but she doesn't. She just feels numb. She goes to find Stuart. The two of them just stand next to one another, not knowing what to say.

One of the helpful things about having animals and children in a crisis is that, no matter how much you want to crawl to your bedroom, shut the curtains, climb under the duvet and succumb to blackness, you can't. You have to keep going. You have to keep feeding and cleaning and loving. And so they do.

On the day after what was going to be their most profitable day of the year, the whole country shuts down. Full lockdown. By law, no one is allowed to leave their house, a sentence that previously belonged in a dystopian novel. The only exception is to allow people to go shopping for essential items or to exercise. Pictures begin to flash up on Rebecca's social media feed. Shoppers with trolleys full of toilet roll and nappies. People fighting over pasta packets. Queues that stretch around the block. No one can get anything they want. For quite possibly the first time in their lives people are thinking, *But where does it all come from, anyway?*

Rebecca looks at the pictures: at the fear, and the shock, and wonders what they should do.

Their farm shop usually opens at 8 a.m. Rebecca

knows something strange is happening when she walks from their house through the yard, past the now empty pig pens, and round the corner to the shop. From where she stands all she can see is people: a great snaking line of them.

'Hello,' she says, and waves a little awkwardly as she goes to unlock the shop door. There is a woman at the front of the queue she has never seen before. She looks at Rebecca with her face half hidden by a disposable blue mask.

'I've heard you sell bread. Have you got some? Do you have any bread?'

'Yes, of course!' Rebecca says. The woman immediately bursts into tears. She cries with relief: an expulsion of fear and shock and understanding that, for the first time in our privileged Western lives, we cannot have what we want, and what we want is food.

Rebecca starts to limit the number of people allowed into the shop at one time. This makes it easy for her to see the weight of people's feelings. Many of them cry when she tells them she does not just have bread but also yeast, and meat, and toilet roll, and vegetables, and fruit and, of course, milk. Milk from cows she knows by name. There are some who are struck dumb by this concept: that what they will eat or drink is made from animals who have lived, or do live, in the field next door.

It has taken just five years for Rebecca and Stuart to change it all. They have felt the weight of the risk they are taking every single day. It has been so relentless that the first time they really stop and tell their story, from beginning to end, is the day I first visit. As we talk and walk through fields of grazing cows they describe the pain and challenges they have faced over the last decade. Illness, death, financial ruin, change, toil, hope. The story fills Rebecca's eyes with tears, and mine too.

They have done so much in so short a period I understand why it feels overwhelming to now stop, stand back and look at it. Partly it is the sheer fact of almost never having a day off. But part of it is also, I think, being able to see the shadow of the other path – the one they nearly walked down – and knowing what it would have meant had they done so.

This land – the home of Rebecca and Stuart – has been a place of farming revolution before. Two hundred and seventy years ago, it was here that the Norfolk four-course system of farming was made popular in England.[7] The change in the rotations of crops prevented the exhaustion of the soil and became a hallmark of Norfolk husbandry. It was this agricultural revolution that enabled the population to continue to grow when previously – during the Roman period, then in 1300 and again in 1650 – it had been unable to.[8]

The Norfolk four-course system became common on farms all over the country. It more or less remained standard practice until the 'green revolution' came along, with chemicals to take its place. Now, there is a new revolution happening here. This farming revolution is about keeping the earth covered, about a diversity of plants and animals and people, about mob-grazing and direct selling. But it is also about something else: compassion, community, connection and learning.

As I leave their farm for the last time, I drive away from Rebecca and Stuart down a road banked by endless flat fields as the late-summer light hits my window. It is that special kind of Norfolk light, shimmering and horizontal, the kind that makes everything half-magical, as though you might turn a corner and find a scene from a time long since passed. And I realize that this too is what their farm is about. It is not just about revolution. It's about the cracks that finally let the light in.

*I think what happens is, you end up coming back to the*
*fact that you need stock.*
*To keep the fertility up, the grazing animals is the biggest*
*plus you'll find.*
*You never grow such good crops as when it comes off after*
*stock's been on it.*

Paul, farmer

# 9. Year Four – Examining Roots

When the new year begins in London it is not just the sheer number of people I notice. In the years since I have been away the gap between city and countryside feels like it's grown even wider. I go to a birthday dinner where the host apologizes for serving meat, acknowledging that such a thing has now become 'controversial'. I start to notice vegan posters on billboards and London Underground trains. After a while I take photographs of them and collect them in an album on my phone. The posters are capitalizing on Veganuary, a movement in January that encourages people to eat only vegan food for the month and, preferably, much longer. Global fast-food brands promise me ecological salvation if I choose their mass-produced vegetable alternatives, grown somewhere (but probably not here) in ways economical enough to make them cheaper still than my Tube fare, rather than a pasture-fed steak from down the road. *The cowpat has hit the fan* the posters laugh conspiratorially. *Don't be a carnibore.*

I have come to think that when chicken no longer looks or tastes like chicken because it is so processed and deep fried that it's really just a vehicle for sauce and breadcrumbs, then it might as well be a vegetable pretending to be processed chicken pretending to be chicken. But does buying it make me a better person?

I sit on the train, looking at the adverts, then close my eyes and picture the land of a vegetable farmer I once visited. He farms 3,500 acres of irrigated land on prime growing ground. One hot spring day I stand in the middle of one of his huge, flat fields and watch a machine attached to the back of a large tractor sieve his soil. It takes out any stones or bumps which may cause the carrots due to be planted here to grow wonky, so they are not rejected by the supermarket who will later sell them. Of course, this also means that any organic matter is sieved away too. The remaining soil is so fine I am able to sink my hand into it like it is black sand, right up to my forearm. I bring up a handful, open my fist and watch it blow away in the wind across the huge flat field. This soil has almost nothing left in it. Everything the crop will need must be added. Growing in it has become a kind of hydroponics.

Statistically, over a third of the carrots grown in this field will be rejected, for the supermarkets think we only like them uniform: straight, medium sized, orange. Sometimes they will be rejected because there are simply too many of them. Later that day, he and I find a whole field of carrots unharvested, left under straw. We dig some up, to see if there is anything wrong with them. There isn't. We wonder whether the contractor who farms these fields simply had met their supermarket quota and could not find anyone else to buy these rejects. When a 1kg bag of carrots sells at 43p, it's easy to see why the cost of getting them out isn't worth it if you've no one to sell them to. Instead they have been left to rot, a whole field of them. As we drive away it is hard not to feel sick at the waste of it all.

We go to see the one field with a slope to it. Last winter the rain poured through the sieved soil and, with nothing to hold it, created cracks so deep that they now look like a dried-out ravine cutting through the field. They are deep enough to reach my knee when I step inside them. We go to field after field to try to find a worm, for his farm manager is adamant you don't get worms in this kind of soil. It is too fine, too sandy, too free-draining. Good for vegetables; bad for worms. Three fields in and I am beginning to think this is some kind of biological anomaly and that maybe the farm manager is right. But the farm has a few old hedges left and good thick grass margins next door to wide strips of winter bird-food cover.

'Can we dig there?' I ask, and point to one of the margins near a hedge.

'Sure,' he indulges me.

We dig one spade's depth, about half a metre from the hedge. In the spadeful of earth are three worms. As I hold the clod I realize that, once, it would have been the same as the earth in the worm-less field next to me.

Now I sit on the Tube looking at the brightly coloured adverts and think of these huge lifeless vegetable fields. I think of how empty of biodiversity they are, big-sky landscapes that are just as much of a factory as any huge animal barn. I think of all the killing that can happen in plant farming even if what is being killed is not intentionally slaughtered, nor furry but instead is microscopic or slimy or shelled or beaked. Then I think of the farms I have been to where farmers have talked to me of 'the golden hoof' and where grazed animals have brought back life – birdlife, wildlife, insect life, floral life, microbial life, fungal life – into a field and its soil. I think how these animals work together in cyclical grazing patterns that fertilize the ground, increasing its biology and its ability to clean air, hold water, sequester carbon and transport nutrients through the plant and into the animals that eat them and, in turn, when we eat the animal, into us.

I find myself standing in a London supermarket, in one of many huge aisles filled with every kind of food I could want, vacuum packed and brightly coloured. I realize I miss my village shops, which still seemed to have everything I needed even though they were a fraction of the size of this one. I close my eyes in the chill of the freezer section and try to imagine who has grown this food, and how, and where, and wonder how far it has come and what percentage of its price they will get for their work. I try to think of the human stories behind each bag of frozen chips and pale fleshed pre-plucked chicken and bunch of bananas. I try to think not just of the ecological cost, but the human cost too.

Now I am back in the city, it is not just that I am thinking more about where the food I eat really comes from. It is that I see how connected I am to it. Standing here, making my choices, I am linking myself to the lapwings that rise up from Hampshire stubble fields; to the frogs that spawn in Cumbrian streams; to the hare that runs on East Anglian maize strips. I am connecting myself to lands I shall never step foot in and to the lives of the people I will never meet. What I don't have any idea about is that, in a few months' time, this city – and the rest of the urbanized world – is about to go through the same awakening.

We drive up to Suffolk each weekend, staying with Ben's parents in his childhood home on the other side of the meadow fields from the cottage. We pack pyjamaed children into the car on Friday evening and return on Sunday evening filled up and smoothed out. I start to notice that if we do not go I feel the absence like a physical knot within my body.

There is another reason we must be there so regularly, apart from this new need I have to fill my head and heart with green. This is the year that we will change the crop fields.

The three crop fields are very different from the meadow fields. They are three slabs of land separated by two hedges and a pathway, set on a slight slope and bordered on two sides by a road and, on the other, by more large fields. This is arable country and the soil is good, Grade 2 'Beccles Hanslope', a fine loamy earth that sits over clay. It is good for growing crops but can be heavy, hard to work and seasonally waterlogged. It is, I will later be told, 'man's land'.

The crop fields are also not right outside the window. To visit them we must travel a short distance. That

Saturday in January we decide to go by bicycle, snatching a few hours between showers of rain, which seem to have fallen almost constantly for the whole of the month. This time we make the children come with us because I want to try to balance their new city life with their old country one, but the journey is terrifying. Both can ride bikes well enough, but the twisting lanes are narrow, with deep puddles at their edges, and Land Rovers and cars fly past frequently enough for me to yell 'WATCH OUT!' every few minutes. It reminds me of the ironic truth that moving to the countryside for green space mostly means discovering you spend more time driving a car than you ever did in a town. I am relieved when we turn off the road onto the path that leads to the crop fields, then realize we will almost immediately have to make the return journey before winter darkness falls.

We push our bikes up the incline of the path that bisects the three crop fields. At its base, on the roadside end, are a smattering of cottages. At its very top is the gate to the house that once belonged to the fields but which was sold off long ago, now smartly remodelled to erase its farming provenance. The children are tired and cold and we abandon their bikes on the edge of the path so Ben and I can sit them on our saddles and push them up to the top of the slope. To our left is the first of the three fields and, on our right, a thick hedge which runs halfway up the path, separating the first field from its sprawling neighbour on the right. We stop at the top of the slope where the hedge ends and the path bends right then almost immediately left before straightening and carrying on to the gate of the old farmhouse. To the left of this second half of straight path is the third and final field, separated from the first by a second hedge.

It feels like it might rain again, so we shelter under the old oak tree at the corner of the path's left-hand bend, looking over the big second field. Two more oaks stand in the middle of the field and, running from them to where I am now, is an ancient right of way slicing across the field. There is another right of way that cuts in half the field we have just walked past. No one has ever seen anyone using it, partly because it saves both crops and boots to walk the field's margins instead and, as we will discover, because the exit that takes walkers from field to road is blocked by decades of bramble growth.

The view from beneath this huge oak can be a good one. One of the cottage's residents, Mr Pink, keeps a canvas camping chair up here all year round so he can sit and look across to the other side of the valley. Usually, I can make out the top of the tower belonging to the church where Ben and I have been to weddings and funerals for those we love, the pale stone visible amongst the trees. Now, though, the sky is hung with the threat of rain and my attention taken up with rubbing small cold hands, and I do not look for it.

It is only that evening, when we look up old maps online, that I realize how very different these crop fields would once have looked. Although their boundary shape is the same these three fields would once have been fourteen, each separated from its neighbour by a hedgerow. The satellite picture I am looking at on my laptop for comparison is an old one. It pretends our fields are lush and green, whereas really they are now great gashes of brown earth, the crops planted within them still a secret. As I move the cursor across the old map, working out on the satellite photograph where the lost hedgerow lines would have been, I suddenly realize that I can see their ghosts on the modern satellite image: slightly darker green lines of legacy fertility left behind, hidden on the ground but visible from space.

I see too how the second crop field, the one with the oak trees in it on the right of the path, would once have been divided into seven. Its top-left corner, which still lies just below the old farmhouse, would probably have been a wildflower meadow, kept for the horses that pulled the plough. Later we will learn that the crops there always do a little better than the rest. A kind of earthly muscle memory means the soil still contains meadowland fertility that the rest of the field – ploughed and sprayed and cropped for sixty years or more – has lost.

From this bird's-eye view of history it is easy to see how the crop fields once fanned out around the old farmhouse like a sun. We later learn that the pattern is called an 'ovoid' – a very early medieval grant of land or wood-pasture that had an oval shape to economize on hedging costs. In places the boundary kinks at the edges of the oval, which suggests that the grant of the land pre-dates the local parish, which would, back then, have been only a settlement. This farm, then, is an ancient one.

Now it is up to us to decide what should be done with it. It's not a large amount of land by modern standards, just under two hundred acres, but now the subsidies that have made it financially viable are being withdrawn. What we are sure of is that we do not want to 'rewild' it: to stop farming it and let nature take over. Although I believe we need wild swathes of land in places where growing food is difficult, this is productive farmland. It is important for us to try to make food production and nature coexist. In the UK, we grow just over half of all the food we consume.[1] The rest is imported, sometimes from countries whose climate is less sympathetic than our own. Some of these countries are already suffering the consequences of climate change: floods, fires and famines. We don't need to look too far to find them. A 2018 report by EU auditors warned that the amount of land at risk of becoming desert in Europe alone is twice the size of Portugal. In Spain, one fifth of all land is now at high risk of desertification, as is agricultural land in Italy, Greece and western North America.[2] Importing our food just means exporting our responsibilities. Why should we get to live in heaven by making others live in hell? This is not the choice we want to make.

Instead, we decide to do something different. We will convert the crop fields to organic farming, as we have the pasture fields.[3] It is the best compromise we can think of between growing food and protecting nature, trying to make the two work together. Our crops will now no longer be grown using pesticides or artificial nitrogen. This year's harvest will be the last to do so.

First, though, we need to break this news to Richard. He has his own farm but contracts to farm others using his machinery and equipment, including these fields which he has farmed for decades. It is a solution that, according to the National Association of Agricultural Contractors, is chosen by 91 per cent of UK farmers who also use a 'contractor'. Richard trained as an agronomist, is experienced, well respected and, for the last forty years, has been able to use all the chemicals available to grow his crops. So now it is not just the land we have to convert. We must convert him too.

I can see it is not an especially attractive offer to make him. Our two hundred acres is tiny compared with the land he already farms, but something in him is clearly curious because he agrees. He will charge us for his contracting services but this will still be a partnership: we will work out where this farm goes together and split both its costs, and its profit, equally. Each of us has something to lose and gain; each of us needs this experiment to work.

The first thing we must do is plan what we will grow, and when we will grow it. Up until now the farm – like many farms – has grown three crops every year: wheat, barley and oil seed rape. Organic farming is different. Each harvest leaves a legacy behind it – of fertility, pests or disease – and so we must find a way to work with that inheritance but without chemicals. Fertilizer must come from the muck of animals grazed on organic pasture and not from a bag. Nitrogen, which is essential to enable crops to grow, cannot be synthetic and sprayed but must come both from manure and from planting legumes, like beans or peas, which have nodules on their roots and can – unlike cereal crops – take nitrogen from the air and fix it into the soil for the following crop to use to grow.

But to grow a greater number of crops at one time we must divide our fields up. The old hedgerow lines, the oaks in the big field and the rights of way make this an easy decision. We print out the map of the farm and draw three new hedgerows down it, laying them down ancient Bronze Age boundary lines, so that in total we will plant two kilometres of hedge and ninety new trees. The two lonely oaks in the middle of the second field will be joined up by one of the new hedgerows, following the line of one taken out five decades ago, which will support the trees' growth and stop machinery coming so close that it compacts and slices their roots. On either side of the hedgerow will run a five-metre-wide grass margin. One of these will reclaim the ancient footpath that few now use.

I shop online for new footpath signs in anticipation of all these new walkers. The following weekend we go with cutters and gloves and clear the brambles from the roadside right-of-way exit. We collect dozens of cans and packets thrown from car windows, fly tipped rubbish and multiple deflated helium balloons, which somehow always seem to find their way onto every farm. Underneath all the growth we find a bridge that crosses a ditch and, although we still never see anyone use it, part of me hopes that one day we might.

*

When we meet Richard up at the crop fields the following month the worry of what we are making him do is etched all over his face. He hops from one foot to the other when he warns of the weeds that will come. He has fever dreams of bur chervil, whose leaves look like its carrot cousin. The weed has taken on a kind of personal malevolence to him and he dreads finding it. He is worried we do not really understand how very hard this will be. He is probably right.

This year will be our last conventional year of wheat, barley and oil seed rape. Thereafter we will grow four crops over five years in a rotation that John Pawsey, our neighbouring organic farmer, has helped us create. We will grow oats, grass, barley, beans. It will, I later learn, be the first time oats are grown on this farm for over half a century.

We go on a trip to visit John, who has been farming organically for decades. I wonder if it is, in part, to prove to ourselves that this thing that so many seem to think is fantasy – growing commodity crops without any synthetic inputs – can actually be real. John is a fourth-generation farmer who stopped using chemicals on his land over twenty years ago, but his is no *Good Life* smallholding. The farm is 1,600 acres and he is the contractor for an additional 2,261 acres of other land alongside it. There is not a hint of hippy to him. He wears an Apple Watch, Diesel jeans and the kind of black-framed glasses that would look at home in Soho House.

John teaches us what the old farmers knew. The key, he says, is diversity. Grasp that and the rest will follow. As we walk John's farm, Richard watches John and I watch Richard. When the afternoon is over they shake hands. Richard thanks him and says that, when John first became an organic farmer, he thought – in fact, everybody thought – he was completely mad.

'And now look,' he says. 'Here we all are, asking you to share your secrets.'

As we walk back into the farmyard and past a row of machinery, Richard, on instinct, wanders over for a closer look.

'My new inter-row hoe,' John says, waving a hand at the machine.

Richard's face lights up. 'Tell you what,' he says, gazing at it. 'We should probably get one of those.'

As we drive home from John's farm, the afternoon is turning towards darkness. I think of the maps we have planned all this out on, and I think of the legacy fertility from the old meadow field and the faint hedgerow lines. I wonder whether, if the earth can hang on to its past after all these years, maybe I can too. Maybe if I stand back and look at myself from above, I will be able to see echoes of a different inheritance, passed down from my grandparents, a shared DNA from a life on the land. And, if so, maybe it will be enough now to replant a new life along the faded lines of the one I left behind.

When March comes, the world stops. The news bulletins flash with statistics and pictures of those who have lost their lives to Covid-19, horror stories of full hospital wards and death and economic catastrophe. The world has become a disaster movie. We become strangely used to hand gel, taking temperatures, not touching, sneezing into elbows, faces hidden by masks, parents turned into teachers.

The queue for the supermarket is so long it takes Ben an hour to get in. He sends me a video of aisle after aisle after aisle with no food on the shelves. His camera phone catches a man in his eighties standing with a tartan wheeled bag in one hand, his shopping list in the other and a look of pure fear on his face. My WhatsApp groups ping a stream of panic. There are screenshots of online queues for food deliveries: *You are number 947 in the queue, your wait time is four hours thirty-eight minutes.* A few days later the panic turns into advice. Links are sent through from wholesalers who've lost all their business and are now delivering not to restaurants, because they have all closed, but directly to customers. I hear about farmer after farmer who has cows out grazing in their fields destined for beef contracts with restaurants that have just been cancelled. I hear about dairies pouring uncollected milk away as people in the cities go from shop to shop trying to find a pint of it. I read about supermarkets filling the gaps in their shelves with cheap imported meat reared in ways our own farmers are prohibited from using, but no one really cares when they are frightened they might starve.

Outside our front door the city is like a film set. The streets are empty of cars, the sky empty of aeroplanes. I read of nature reclaiming cities; of birds visiting gardens in numbers not seen for as long as people can remember. In the news, death is everywhere. Outside, the spring shows off with sunshine and blossom, as though to assure us there is still life and beauty left in the world.

But it feels like this turning of the seasons from winter into spring is happening without me. I am locked in a city away from it all. The playgrounds are shut, the swings and slides draped in fluttering yellow tape like a crime scene. The paths are full of joggers and dog walkers anxious to keep their space. I spend my walks with the children circling the same park. I remember that city grass is different from country grass, shallow and exhausted, bruised by a thousand footsteps. Missing the spring because we cannot go to the farm makes me feel unmoored in a way I had not expected. I forget the day or week or sometimes the month. I had not properly understood how the ability to measure the time of year by what was outside my window had anchored me to the world.

I think of those in hospitals – patients in beds and medical staff leaning over them in layers of personal protective equipment – and of the study that looked at the recovery of patients in a Pennsylvania hospital between 1972 and 1981. One set of rooms in the hospital had views of deciduous trees, the others a brick wall. Patients undergoing the same operation were assigned rooms randomly. Those with rooms that looked out onto the trees stayed in hospital for shorter periods, needed fewer painkillers and suffered fewer complications afterwards.[4] Nature can heal, even if you are just looking at it through a window. What, I wonder, are our Covid patients looking onto? What are the nurses and doctors treating them staring at all day? I think I know the answer, at least in the city in which I live.

Schools eventually return and lockdown is loosened. This year is full of plans for organic conversion and Countryside Stewardship applications. We have applied, and been granted, extensions for both, so as soon as we are allowed we go back to Suffolk. For now we cannot stay with Ben's parents, and so we cram into my old writing hut in one of the meadow fields, arriving in darkness after months away. I climb out of the car then reach to lift Wilfred from his car seat into the night air. When I straighten I realize, with a kind of shock, that I had forgotten what the night sky can look like. I stand for a moment with the heavy warmth of a sleeping child in my arms and the sweet, sharp smell of night, while above me stars sweep across the blackness of the sky from horizon to horizon.

The dawn chorus wakes us just after 4 a.m. The children burst from their beds and I open the hut door to let them out. The meadow smells sweet, the dew wet between the grasses and wildflowers. They stand in the dawn and pee small steaming streams into the tall grass under a pink sky. Then they run and run and run. They cry out and chase one another around and around, until the legs of their pyjamas are soaked through with dew.

The meadow has so many buttercups that, from a distance, it shimmers yellow. The new-born lambs the boys had held in spring, bleating and shrill, are now turned out into the neighbouring field. I watch them as they mob up to form small gangs, leaving their mothers and tearing from one side of the field to the other like children in a playground. In a few months' time they will stop skipping and jumping and put their heads down to graze, and I will miss seeing them play.

I want to spend our journeys looking out of the window because everything is blooming so fast and full that it feels as though the trees and hedgerows are in competition to see who can put on the better show. The lanes are filled with avenues of cow parsley. Verges are lined with pale yellow cowslips hanging from green cases in ready-made bunches.

To begin with, I think I am seeing wildflowers everywhere because it has been so very long since I've seen any at all. Then I wonder if it's because I now know their names so I notice them more often. But, after a while, I am sure there are just more of them. Road verges that would once have been mown are a riot of colour: they are lined with poppies, cornflowers, buttercups, daisies and the milky white of cow-parsley. There are more primroses, hanging heavily clustered in their little green cups, than I have ever seen before. I wonder if they, like oaks, can somehow have a mast year, a bumper crop to ensure their survival. It is hard not to think that, somehow, the world knows we are all grieving for a life we have lost and is trying to show us she can still be beautiful.

I find out that these verges are an accidental outcome of austerity. Council budget cuts meant they could no longer mow verges regularly to keep them 'nice and tidy'. As they were only now cutting once or twice a year, the cuttings were so tall they had to take them away rather than leave them to mulch down as they used to. Taking the grass cuttings away reduced, rather than added to, the ground's fertility. Now the seeds of wild-flowers, a bank of which had been dormant in the soil of the verges for decades, finally had the right conditions to grow. And they have.

Plantlife, a wildflower conservation charity that has been campaigning for the changes in management of verges since 2013, claims that 700 species – around 45 per cent of all native flowers – have been found on our road verges, which now constitute 1.2 per cent of land in Britain.[5] In Dorset, where the annual budget for highway verge management has dropped from nearly £1 million to £650,000 in five years, the verges of the four-and-a-half-mile Weymouth relief road have been visited by more than half of all the species of butterfly known in Britain, many of them now rare.[6]

In a year marked by disease, death and loss, I find myself looking for its counterweight. This year it is easy to find, and not just on the verges I drive past. I read about three new meadows in Gwynedd, north Wales, one of which has five times more species of wildflower just one year after it was restored. In Norfolk, seeds from a wildflower reserve are used to recreate a meadow and, the following year, a rare sulphur clover is found that had not been part of the original mix. Nature, given the chance, recovers fast.

Later, we take the children up to the crop fields. By the time we get there the sun is strong. The oil seed rape is thick and acid yellow. I remember when I first began to see the colour in the fields on the drive to my grand-parents' farm. Someone once told me that the yellow of rape is so bright that, in some people, it can induce fits.

Farmers only began to grow oil seed rape in the 1970s to give the soil a break from cereal crops. Now this rape will sell for more than the wheat and barley. Just before it is harvested, like other crops such as wheat or beans, it is desiccated with glyphosate. Even though the herbicide wasn't initially created or marketed with this in mind, some farmers in the 1980s realized that spraying glyphosate just before harvest would, in killing the crop, reduce moisture levels so that the cost of drying was saved, and meant harvest could happen weeks earlier than if the sun were made to do all the work.[7]

This is the last harvest that so many fields will be covered in the crop's yellow. Oil seed rape has been dev-astated by the cabbage-stem flea beetle since a ban on coating rape seeds with neonicotinoids, which killed the beetle when it laid its larvae in the newly grown stem. The chemical was found to fatally harm bees and other insects, but even so, since 2016, the EU has granted 205 emergency derogations to other countries so they can continue to use seeds dressed in this powerful insecticide.[8] When the UK rape crop fails, the seed or oil is just imported from a country allowed to use the chemical. I hear of one reckless farmer made so angry by the hypocrisy that he attempts to mix his own 'neonics' himself.

The boys run in and out of the rape, their bare legs whipped by the prickled fan of their leaves, the plants tall enough for them to hide in. When I persuade them out their faces are freckled by the pollen, as though someone has tried to make a leopard of them with a felt tip, the yellow so bright it looks artificial. This will be the last time we will grow it. Next year this field will have oats in it. As I look at the garish yellow covering the valley, stretching on for miles, I realize I am glad.

'Read out the guidance again, would you?' Ben says.

It is the summer holidays and we have left the children with their grandparents and are making our way to the crop fields alone. With just us two in the car it feels almost like a date. In the end it will turn out to be romantic, just not in the way I had expected.

As we drive I notice I am starting to pick up the farmer's habit of weighing up other fields. I peer out of the window, crane over hedges, turn back in my seat to get a better look. I want to see who has planted what – 'Hey, look, he's brave to have another go at oil seed rape' – and who hasn't – 'Didn't manage to get anything in over the wet autumn, then'. I notice who has taken their field edges out of production and sown them with wildflowers, who is planting trees, who has interspersed his largest field with big fat strips of sunflowers, who has sown their crop into a vast hedge-free field right up to the edge, who has excavated their roadside ditch so it is deep and cavernous. I understand now why some experimental farmers tell me they grow their hedges high to block out other farmers' judgement. I think of those who have told me of farmers who spray an extra two tramlines of artificial nitrogen on the crops nearest the road to ensure these grow the tallest. I remember the land agent who showed me a picture of a sign a contractor put in his field saying 'CONSERVATION AREA: LOW INPUT CEREAL' in case his neighbours thought his crops poor, even though he was being paid extra to grow them this way. I think of something a young farmer once told me about his late father and his friends: 'Their idea of the perfect day out,' he laughed, 'is a right good nosy around someone else's farm.' Now I sort of understand what he means.

But on this journey I cannot look out of the window. I have to read out instructions to Ben. This time it is for a Farm Environment Record Map that we must complete and send off with our Countryside Stewardship application.

'OK. So. We must record . . . "all environmental features and areas on the land parcels to be included in the application and mark parcels that are at moderate to high risk of runoff or soil erosion".'

'OK.'

'I don't even know what that means. Right. One of the "environmental features" is trees. Another is hedgerows. OK, listen to this.' I read from the screen on my phone. 'Only boundaries with, on average, one or more eligible tree for every hundred metres need to be marked as "boundaries with trees". For example, a hedgerow of 400 metres would need to have at least four eligible trees along its length. Eligible trees are those that are native species, standing within one metre of a hedgerow and over thirty centimetres diameter at breast height.'

'Yeah,' said Ben, turning up the farm track to the fields. 'Sounds about right.'

We walk the boundaries of the fields measuring tree trunks and recording the ones that qualify onto a map. We have been walking for well over an hour when we move across into a triplet of little fields alongside ours, which a neighbour has asked us to farm for them. We walk along the hedgerow and are nearly at its end when we see a cluster of five very tall trees, set back a little, their canopy high and deep. Ben, who knows more about trees than I do, frowns. Out of the two of us, I am more typical. In a 2013 YouGov poll, conducted for the Woodland Trust, more than four-fifths of people couldn't identify an ash tree from its leaves and nearly half could not recognize an oak.[9]

'That's weird?' Ben says.

'What . . . ?'

'I can't work out what they are.'

We walk up to their trunks, which are tall and lean and stately.

'What is it?' I ask him.

'I honestly don't know. They look a bit like . . . but they can't be . . .' He tails off, moving towards one, touching its bark and looking upwards. 'Does that plant app you've got work on trees?'

'Yup.'

'Can you try it on these?'

I go and stand next to him and reach up to pull at one of the ends of the branches, photographing the leaf. The app flashes – *identifying* – then a page slides up.

*FIELD ELM*

*Botanical name: Ulmus Minor*

*Common name: English elm, common elm, Horse may, Atinian elm, Small-leaved elm.*

*Genus: Elms*

Ben looks at the tree, his mouth hanging open slightly.

'*Ulmus minor*,' he says quietly, as though he were greeting it. Then he walks backwards into the field, taking in the height of it. 'It's escaped, somehow. But how? I've never seen one so huge?'

He stands looking at it then says, quietly, 'I never, ever thought I'd get to see one like this.' When I turn to look at him I see his eyes are filled with tears.

Later I will realize I have seen trees like this before. A few months earlier my second cousin, Rob, sent me an email with a file attached. When I opened it I saw it was a digitized copy of an old cinefilm, taken in the 1940s. I clicked *Play* and my grandmother flickered onto the screen. She was so immediately and utterly real on the laptop in front of me that my throat tightened and I laughed at the joy of seeing her alive again. Her nineteen-year-old ghost moved in bright technicolour, dressed in a gingham dress and pigtails. She was standing in farm-land next to a huge, billowing hedge, plucking blackberries from it, sucking the juice from the tips of her fingers. Then she walked towards the camera down a yellow brick path in a scene that looked like an oil painting. The fields and hedges and skyline around her looked both familiar and also strange, and I realized it was because of the trees – inconceivably tall and full – which lined the edges of the field alongside the fat hedges. The shape of them made the landscape seem somehow foreign. I watched my dead grandmother walk unthinkingly into a field filled with a horizon of trees that I have never seen and could not name. They must have been so common-place to her that she surely believed they would be there forever, for elms were once an emblem of England, native descendants of the wildwood that covered much of the country 8,000 years ago. They were able to grow hundreds of feet high and twice as old in years, with a canopy that could stretch over a quarter of an acre. Because of this, they were often planted to mark the way on old country paths as they grew into a living road sign, tall enough to be seen above mist.

The cinefilm of my grandmother was taken around twenty years after Dutch elm disease first appeared in the United Kingdom in the 1920s. Since then over 25 million elm trees, 95 per cent of all mature elms in the United Kingdom, have been destroyed. It arrives in the form of a fungus spread both by a bark beetle and from tree to tree via their roots, the first signs of which are yellowing of the leaves in the summer that leads to the drying out of branches. Next comes a blackish brown stain that spreads on the fresh wood beneath the bark and then, when more than half of the tree is consumed by the disease, it tries to kill itself: plugging its own tissue and preventing water and nutrients travelling up the trunk. Young elms will regenerate from the roots of the dead elms, or from their seeds, but they rarely survive more than fifteen to twenty years.

Ben and I stand next to the trees and take photos of each other gazing up into the elms' canopies as though they are a tourist attraction, which, for Ben, they sort of are. Later we discover that neither Richard nor the land-owner had known the elms were there. For years people must have walked unthinkingly past them, as we had. But they were here all along, hiding in plain sight.

I later read of the elms' symbolism in literature and folklore: planted by nymphs on the tombs of Kings and heroes in the *Iliad*, symbols of pastoral life that cast shade which becomes a place of special peace in third-century poetry, and the gateway between this world and the one after. As Ben and I walk back to the car he talks about the trees, then takes my hand and I think that maybe this new kind of farm-bureaucracy-date might be all right after all.

I am eight weeks pregnant with our third child on the day I touch the electric fence. I catch it with my elbow, lean-ing over to hold out my hand, trying to coax over the new calves in the meadow field. The wire is up to keep the cows, their calves and the gentle bull away from the entrance of my writing hut, which opens straight into the field. Farmer Martin only put it up to keep us safe.

It is the slightest of shocks, enough to jerk my arm upwards. The calf with his nose nearly in my palm jumps back in fright. It is not the same kind of shock that Aubrey once got, taking the wire with a firm, fat grip, which then ran through his small body so hard that he turned with a face full of hurt and shouted at his brother, thinking he had thrown a ball at his back. For me it is light and sharp, like its name: *a shock*.

Still, that evening I Google it: *electric shock pregnant.*

*The uterus and amniotic fluid are thought to be excellent conductors of electric current which reaches the foetus causing cardiac arrest and foetal death.*

It warns that the data is skewed. Women who have an electric shock and go on to have a healthy pregnancy and baby rarely think to mention it. I text Farmer Martin anyway.

> Martin, do you happen to know the amperage of the fence?
> Stupidly touched it. Am 8 weeks pregnant.
> Apparently the two don't go that well together.

It is late, after 9.30 p.m., and the sun is preparing to sink, but there he is, with a small brown ammeter and a worried grin. He waves away my apologies for bringing him out on a Friday night. A farmer has no working hours. Nature and its cycles happen on their own clock and not ours. If you are needed, you are needed, and so you go. I am lost in the talk of double currents and wattage. The fence is the lowest strength it can be, well within regulations. He is sure it will be fine. I say I am sure it will be fine too.

I start bleeding just over two weeks later. I call the emergency line the midwife gave me but they tell me it is probably just pregnancy spotting. In my stomach I feel the low drag of cramps. I know it is not spotting. The pandemic has restricted antenatal units so I pay £130 for the internal examination that confirms the foetus is dead. The sonographer, her voice kind through her Covid mask, tells me I am clearly in tune with my body and so the miscarriage should begin quickly. When I ask her what miscarriage will feel like she says it will feel a little like labour. The cervix will need to open. It will hurt. I am grateful for her honesty and am glad that she does not say, as I have read, that it is like a 'bad period'. It will turn out to be almost exactly like giving birth, except with much more blood and no baby at the end, just emptiness instead.

As I wait for the miscarriage to begin I long to be walking in the meadow field. All I want is to put my bare feet in the grass and be allowed to be an animal, labouring a scrap of baby into the earth and burying it there. Instead I go alone to the greenest space I can. Clapham Common is full of ice-cream vans and picnickers and joggers. I try not to look at anyone but walk instead towards the copse at the end of the common. I try not to cry in case, I suppose, it freaks someone out and they think me mad. Amongst the paths fringed with Coke cans and polystyrene food boxes and crisp packets with soil worn down to dust, and amidst the sirens and horns and air which tastes of fumes, I notice a cloud of tiny brown moths rise up as I walk through a patch of long grass. I stand in the grass for a long time and watch them scatter up towards the sky. Then I go into the small copse and sit under the oak at its centre. I think about all the death and life I have seen this year, all the cycles, the darkness and light, and how I have been changed by them. The sun falls through the leaves onto my face and my mind quietens. I close my eyes and try to imagine I'm sitting surrounded by the hum of crickets rather than exhausts and engines, back in the meadow, where I may or may not have accidentally killed my baby.

Summer begins with thunderstorms, wind and heavy rain. On the day that Ben and I stand in the middle of our crop field, though, the day is clear and hot. The summer sky is bright blue, hung with clouds so perfect they could have been borrowed from a child's mobile. It is an important day. After all the planning, this is the first day these fields, technically, convert to organic. From now on we will be unable to use any artificial nitrogen or pesticides to grow the crops we want to sell.

A few years ago I would have noticed only the uniform yellow of the wheat now around me and felt pleased that I could tell its kernelled ears from the wispy ones of barley. Now I realize how differently I'm looking at it. I brush my hands along the wheat as I walk through it but this time I am looking at the ground. I see deep cracks and fissures in the dry soil and patches where the crop has failed to grow. I notice that although the grassy margins at the edges of the field are wide and full, the only life within them is a handful of cabbage white butterflies, bouncing between some purple thistles. There is no hum of insects. The air is almost silent.

We are here to work out how much of this land we should stop farming and where we might sow wildflowers instead of wheat. As we walk to each field edge, I realize the decision is going to be easier than I had thought. In every single field corner the crop has failed to grow and the soil is almost bare, a big triangle of it with just cracks and weeds to show for all the work. It makes no sense to spend money ploughing this part up, buying crop seeds to sow then trying to fertilize and spray some kind of life into the soil. These gashes of earth seem to have nothing left to give. I rest the field map on Ben's back and, with a red felt tip, draw a line that will mark the new crop boundary. We take huge chunks of the field out of farming.

As we walk around the field's edge I see something else. Peppered within the crop are darker patches. Long, cylindrical spikelets of another kind of plant, standing tall and thin within the wheat. Later I find out its country name is 'slender meadow foxtail': a pretty name that suits it, for this is exactly what it looks like, although the farmers here call it something which sounds much darker: 'blackgrass'. This plant has become the nemesis of English arable farmers. Its roots compete with the wheat for nutrients, drawing for itself what should have gone to the crop. Once the wheat is weakened this cuckoo plant sees the chance to take the space for itself. Each plant can produce thousands of seeds, which drop to the ground before the crop is ready to be harvested, so that when the ground is prepared for a new autumn crop its seeds are all there, ready to grow the following harvest. Enough of them can reduce a crop by 70 per cent.

Its spikelets poke out, waving cheerfully, as though it is mocking us. It should. This field has been sprayed with glyphosate twice to try to get rid of the weed. In a last attempt the areas where the blackgrass was worst were doused a third time. It hasn't worked. Blackgrass has, over the last two decades, become resistant to what once killed it and now shakes the chemicals off like a rain shower. So, next week, we will pay five people to walk in a row across the field hand-rogueing – pulling the plants out by hand one by one, like the men who once did the same for wild oats in my grandfather's fields.

Afterwards we walk back to the path that runs alongside the field. Against one of the hedge trees, I see an old spade that has been left there. I don't know whose it is, but it feels like an invitation. On impulse, we take it into the field. We choose a bare patch in the lower third, wary of disrupting any crop. Ben digs: one spade, two spades, three spades deep. We want to see the soil. I don't really know what I'm looking for particularly, only that I know I need to look for earthworms. I must look for the little skinny ones called 'epigeic' worms, which live near the surface of the soil feeding on dropped plant matter and turning it into topsoil. Then there are 'endogeic' worms, which live lower down, creating a web of horizontal burrows. Then, deep beneath the surface, lives another kind: 'anecic' worms. These red- or black-headed worms make permanent vertical passages to be their homes and can live for up to ten years. The largest in the UK was found in 2016 and measured over 40cm. The people who found it called it Dave.

I have learned that, as I walk unknowingly around, worms are performing a magic trick beneath my feet. Their digestive systems concentrate the organic and mineral elements of the rotting leaves and organic matter they eat, so that their casts – their excrement – is far richer in nitrogen and nutrients than the soil around them. Casts left on the surface rebuild topsoil. In the right conditions they can form a layer of nutrient-rich topsoil 5mm deep in just one year. One New Zealand trial found that, over three decades, worms were able to build an 18cm-thick topsoil. Another experiment added worms to a worm-free permanent pasture and measured an increase in pasture growth of 70–80 per cent.[10]

Within the soil dead worms decompose rapidly, adding nitrogen for plants to use to grow. The tunnels they create in their underground world allow plant roots to stretch deeper and reach extra moisture and nutrients. The worms' burrowing loosens and aerates the soil, which means it drains better. When rain falls heavily, a field with many earthworms will drain up to ten times faster than one without. I have also discovered that deep ploughing reduces earthworm numbers. Researchers found that fields that had not been ploughed for four years had twice as many worms as annually cultivated soils.[11]

I will later learn that midsummer is a terrible time of year to be playing at worm hunts. The months of dry weather may well have driven the worms to burrow deep into the ground in search of moisture. Still, knowing none of this, Ben digs down, three spades deep. I had read that eight to fifteen worms per spade was a good indicator that there was life in the soil: some innate biology that means it is – as it is supposed to be – alive. We sift through the clods of dirt. In the three spadefuls of soil we find two worms. I pick up a clump of earth and bring it up to my face. In her book *Losing Eden*, Lucy Jones explains that chemists at Brown University in the USA discovered that geosmin, which literally translates as 'earth smell', comes from two parts of a bifunctional enzyme working together. Geosmin affects activity in those areas of the brain connected with relaxation, which might explain the multiple research projects which have linked gardening with improved mental health. Jones writes: 'Humans are acutely sensitive to the smell and can detect low concentrations of geosmin at five parts per trillion.'[12] But the soil in our field doesn't smell at all. It doesn't smell of earth. It doesn't smell of clay. It doesn't smell dry or like the dusty smell that hot summer earth can have when you crumble it between your fingers. It smells of absolutely nothing.

We leave the first field and put the spade back where we found it. Then we walk up the rest of the path with the field on our left and the old hedge to our right. At the top, just round the corner from where the hedge ends, is Mr Pink's camping chair in its usual place looking over the valley. The big second field had oil seed rape in it but this has now been harvested and most of the field is covered in stubble, although there is none in the corner beneath Mr Pink's chair because the crop never managed to grow here. Instead, there are clumps of weed on a wide patch of bare earth, shaped like an island, all edges and fingers. For the first time in many decades this field will now not be sprayed off with glyphosate. The ground will not then be ploughed up ready for an autumn crop to be planted because, this year, there will be no autumn crop. No crop will be planted until spring, to allow whatever weeds are lurking in these fields to show themselves, so we can try and hoe them out.

Not at this field's edge, though. Instead, this will be filled with plants that sound as though they are ingredients in a Victorian recipe book: brown mustard; fodder radish; white millet; linseed; gold of pleasure. At the other field edges we will sow a wildflower mix that sounds like a spell: sainfoin, red campion, musk mallow, yarrow, knapweed, wild carrot, ox-eye daisy, plantain, fescues, wild grasses. Sowing these plants in soil that struggles to give us a crop is not just about the money we will get from Countryside Stewardship to plant them. These plants will push their deep roots into the soil, improving its biology and structure, helping it store rainwater, which, in turn, can then be used by the crop when rain is scarce. The flowers will bring insects to feed off them, some of which will be predatory and will feast off the aphids that eat our crops. But there is another reason I want to grow these plants and wildflowers. They will be beautiful. They will change the way this field looks and feels so that those who pass it stop, and look, and have their hearts and heads lifted with wonder at the beauty of the life within it. For creating a place that inspires awe in those who witness is, I have come to believe, a public good too.

I try to picture how different the fields around me will look with all the changes we are making and it makes me think again about what being a farmer might come to mean. Most farmers, when asked, will say that it is growing food which defines them. But it has not always been this way. Farmers were once responsible for producing materials – wool from flocks of sheep, cotton from plants, leather from animal hides, linen from flax, viscose from wood pulp. All these were essential to their income and the communities who relied upon them. They were worth so much that whole towns and cities were built with wealth earned from their trade. Even now, the House of Lords in London still has a red, squat seat called 'the woolsack' to recognize the wealth that wool, not meat, brought into the country. Farmers too were once responsible for producing wood – hazel or willow to make fences with or the coppicing of wood and hedgerows for the fuel essential to people's lives and homes and work. I wonder, then, if it is right to say that being a farmer once meant producing what people needed from the land: be this food, or fibre, or fuel.

Now, 62 per cent of all English farms have diversified

their business in some way because they cannot make enough money from producing food alone. Now, some farmers cover their land in concrete and sheds and rarely step outside their office. Now, it can cost more to shear a sheep than the price paid for its wool.[13] Now food can be produced in a laboratory petri dish, or from a 3D printer, or in artificially lit plastic boxes in disused underground railway stations, and the people who make it rarely call themselves farmers.

The root of the word 'farmer' cannot help unpick the riddle of what farming is meant to mean. The verb 'to farm' comes from the Anglo-Norman *ferme*, which means 'a rent for land' (as in, 'to farm out'). But from as early as the twelfth century, the meaning of 'farmer' has been closely linked to definitions of stewardship. In the 1300s, the word 'husband' was used to mean 'a man who tills and cultivates the soil, a farmer, a husbandman'. The earlier, Anglo-Norman definition of 'husband' had been 'steward'. There is agreement, says one dictionary, that 'farmer [in] the purely agricultural sense is comparatively modern'.[14]

Now there is a new term to grapple with: 'Natural Capital'. Officially, it means the parts of a farm that provide its bedrock, both for the food grown upon it and for the rest of us. The Environment Agency says that farmers should be paid both for increasing their soil's organic matter and creating habitats that increase biodiversity.[15] Farmers in the future might find themselves making more money from their method of farming than from the food it produces.

But I am starting to realize that, while farming has always been about the care of land, stewardship and husbandry, the truth is that for all the trees and bees we still need farmers to grow our food. With supply chains threatened by politics, pandemics and climate change, this food needs to be as local to us as possible. But in a world where micro-plastics produced by fossil fuels can be found atop the highest mountains and in the deepest sea trenches, we need farmers to grow our fibre too.[16] We need them to grow our trees and hedgerows and plants, not just for wood but also to clean our air and provide homes for some of the creatures that form part of our ecosystem.[17] We need farmers to create habitats for insects, for they are at the bottom of the food chain and support the rest of life, and therefore us. We need them to grow flowers for pollinators because over a third of the world's crops depend upon them.[18] We need them to maintain soil full of fungi and micro-bacterial life because these tiny living things now turn out to be critical for so much of the way our world works, some think even for the weather.[19]

But there is something else we need our farmers for, something that is both intangible and unsellable, and which stitches together the fabric of communities with a thread that goes back hundreds of years, sewing our future to our past. We need them to remain the corner-stones of our countryside for all those who live and work and visit there. We need someone to remember how to lay a hedge or a drystone wall and to leave a hog hole for sheep and small children to escape through. We need someone to push the snowplough in winter and wield the chainsaw after an autumn storm. We need someone to stand in the pulpit at the funeral of the last of a generation and tell the story of what it is to live a quiet and useful life to a congregation with swimming eyes. We need someone to remember the old names and the old ways and teach them to us so that we can know them too, because they rarely find their ways into textbooks or agricultural universities.

I am starting to think that maybe it is not so much that the definition of farming is changing, but that it is reverting. Is a farmer actually someone who uses their land to provide what their community needs: food, fibre and fuel but also oxygen, ecosystems, clean water and soil that is biologically alive enough to work with plants and grazing animals to lock up carbon and cool the planet for us. And is a farmer not also someone who can care for the land so that those who are able might leave the grey concrete of the city behind for a moment, stand on a hillside and feel their hearts tighten as they remember that the world, and much within it, is truly extraordinary.

I stand next to the car and look at the faded blue canvas of Mr Pink's chair. I wonder what this bare patch of earth will look like next year, and the year after, and the year after. I picture Mr Pink sitting here in the spring sunshine, looking at new life. I hope it will bring him joy. If it does, even if nothing else works, this alone will at least have meant something.

*

We stand with Richard looking out over the crop fields. The wheat and barley are nearly ready to be harvested. I can tell that, although we have planned these changes all together, he is faltering. He is proud of what he has grown here after a winter that seemed to want to drown us and a spring that hoped to do the opposite. He is right to be. Some farmers have ploughed their crops back in because so few grew that it was not worth the money harvesting them but Richard will coax 8.6 tonnes of wheat per hectare from this land. The national average this year is 7.2 tonnes.

Ben shows Richard our map, lined in red. As he explains that we want to square up the fields, cutting the corners off to plant environmental options for wildlife, I squint up at Richard and try to read his face. Does he think us fools? Is he frustrated to be asked to do all this by someone who's never driven a combine or drilled a field? Someone who has never battled floods or drought or rooks or slugs or disease, or lain awake at night and worried what each would do for that year's harvest? Who has been doing this for the equivalent of five minutes to his whole life? Why should he pretend we are talking sense when everything he knows tells him our ideas are a city person's version of it and, in the end, he'll be left with the mess, the blame and the cost when it fails?

Richard raises his hand and tugs it through his thick hair, then sweeps it towards the field.

'Looks all right, doesn't it? Yeah. Not bad, not bad at all, considering. And now . . .' he trails off, looking down, kicking his boot against the grassy margin alongside the path. I know what he wants to say. *And now you're going to mess it all up.* As I watch him, the weight of the responsibility of what we are trying to do, of what we are asking him to do, feels suddenly very heavy.

Sometime later we get an email from the farming co-operative group that sells our crops for us. Reading it is surreal. Our fields are now linked to a global market. The price of wheat grown here is determined by the drought in Argentina, the strength of the Ukrainian grain harvest, the long-term reserve in Egypt. It feels big and unwieldy. I wonder how I can really claim to be 'feeding the world' when I'm not even sure what part of the world our little patch of earth is feeding.

We also get the year's accounts. The co-operative has sold the wheat, barley and oil seed rape for a decent price. Then I work out that nearly 40 per cent of this income has already been spent on the artificial fertilizer and sprays used to grow it and most of the rest on Richard's machinery and diesel and labour costs. A whole year's work on this farm, before government subsidies are added to it, has made Richard and us just over a few thousand pounds of profit each. We are not exceptional. This farm is the size of an average farm in Britain and average farms in Britain have been making most of their profit from government subsidies. Even the largest farms get nearly half of their profit from subsidies rather than from the food they produce.[20] Later, when I see the farm accounts of others – farms far bigger and more intensive than us – I realize that breaking even can be a good year.

I begin to think that maybe, for our farm, at this time of increasing costs, biological resistance to sprays and the withdrawal of subsidies, what we are trying to do might not be so stupid after all.

We line up across one of the crop fields – the children, their grandparents and our neighbour Giles – each one of us a human post. We are marking out the new hedgerows that will divide the three fields in half. The hedgerow we are now pretending to be will border the public footpath that runs across the land and which few seem to use, although maybe this will change that.

The season is turning. The sky spits rain and I pull my coat tight at the neck. Richard stands, our conductor in a boiler suit, at the edge of the field. His arms wave semaphore directions as he bellows, his blond-grey hair fanning out in the wind like a lion's mane.

'*Forward, more, more, stop, back, back: there, there, THERE!*'

I drive my flag post into the ground. Tiny fibreglass shards embed themselves in my palm. The other human posts either side of me are too far away to talk to. My phone is out of battery and I fiddle with it in my jacket pocket. Overhead, a buzzard wheels; it is the only bird in the sky. The stubble stands ankle-high and I watch my sons run through it wearing summer shoes, and suddenly remember that feeling from my own childhood, the sharpness of stubble scratching ankle-bone, and feel grateful for the chance to give them this feeling, and this memory too.

Richard, happy with my position, takes to waving and

bellowing at my neighbour. With nothing else to do – no way to check emails, or social media or send messages – I sink down to the soil to look at it, crumbling it in my hand. There are a few husks from the harvested wheat sitting on top, a near translucent pale yellow, the size of half my thumbnail. It seems impossible that anything could have grown from something so small. I am hit by just how magical this is: growing things. In this time when the world feels so politically and climatically unstable and full of uncertainty, and the future feels unknowable, it feels unexpectedly reassuring.

I look at my sons chasing each other in the stubble – bootless, coatless, unbothered by the rain – and think that what I hold in my palm is truly as miraculous as having grown them, actual human beings, inside my body. I think of the tech in my pocket and all the clever things it can do. And I think how strange it is that I now find my phone's tricks commonplace and the ability to grow food from the earth the miracle.

In the meadow field, toadstools have sprung up out of the cow pats. Their spores love the warm, wet autumn and, as the weeks pass, I notice different ones appearing. Hare's foot inkcap, small and droopy. Rings of enormous parasol mushrooms whose flat mottled tops are bigger than both my hands. Meadow mushrooms, smooth and white.

I get the children to dig in the dried cow pats with sticks and look for dung beetles. They flip the dry part over and shriek when they see the beetles, small and black and glossy, burrowing quickly away so that it is hard to see how big they really are until we root them out completely. Afterwards they pick the pats up and strut around the field with them held above their heads like serving trays.

As the season turns to autumn, we begin the first year of our organic crops. I stand alongside the first field and watch as seeds are poured into a hopper on top of a huge red drill, the tubes across its back ready to drop the seeds into the ground. On the map in my hand I have written the label 'grass', but what we are planting here is much more than that. It is a mixture of herbs and clovers that will last all year, the plants' roots stabilizing the soil. This is the field that had the blackgrass in it. This is the field where the soil smells of nothing.

In the third field above this one we have planted a cover crop. Despite its name, it will never be harvested. Its job is literally to cover the earth over winter until the cash crop – the sellable crop – is planted in spring. This blanket of plants will keep in the earth's heat; its roots will hold the soil together, aerating it and allowing any heavy rain over the winter to be absorbed by the plants and kept within the soil below for the crop to use later, rather than washed off the field.

As we leave, a crowd of skylarks take off from the stubble, which would in every other year before now have just been sprayed. I remember reading how once there were so many larks that sometimes the farm men would catch some and take them in a covered cage to the pub where, on a signal, the cover would be taken off and each lark's song timed, the longest being the winner. Why not? It was good sport and there were enough birds to spare a few. Back then there were limitless numbers of creatures: hundreds of starlings to steal grain and shit in the animal feed; thousands of sparrows to erase huge chunks of newly planted crop; hordes of hares and overwintering geese to destroy great swathes of barley. There were just so many of them. Until there weren't. In the skylark's favourite habitat of farmland a switch from spring- to autumn-sown cereals and a switch from hay to silage saw their population fall by 75 per cent between 1972 and 1996.[21] It is still declining. I look into the sky and watch the larks floating tiny and high on thermals above us. They are so loud that their song drowns out the roar of weekend motorbikes on the lane at the bottom of the field.

The following month, November, the Agriculture Act 2020 comes into force. From now on the public subsidies that have been supporting farming for so long are to be phased out. By 2024 they will be halved. By 2027 they will have gone completely. I read articles calling the new law a 'landmark moment' that will 'change British farming for generations to come'. I remember how some farmers told me this would never happen. They were sure our government would never get rid of subsidies for food production, not altogether. But now, they are. I hear of farmers vowing just to become more intensive and ploughing up the grass margins they no longer need to qualify for the remainder of the subsidy. But even the most cynical of them could not have anticipated that our government would then do something that so many people – some of them very important people – told me they would not: sign a huge deal to import food produced in ways it would be illegal to use here from the other side of the world, restriction free. I listen as ministers give interviews saying there is no need for British farmers to worry because the deal gives them enough time to 'scale up' their own food production. And I wonder again how this new exciting wave of farming that I have seen swell and rise over the past few years can compete with this. Will nature now exist only for those who can afford to let it, while the rest of farming 'scales up', creating pockets of heaven for those who have the money to live there and wide swathes of hell for everyone else? Because if this happens, it is not just nature that will fail to thrive. We won't either.

When winter comes, I discover a bat superhighway. Its lane is in the meadow fields, at the top of Sheepwash field, opposite the copse. If I go at dusk and stand still for long enough the bats begin to zip past me, close enough for me to hear their clicks. *One, two, three, four*: tiny darts of black. They pass right to left, then wheel around in the air and come back again as though they are playing with me.

Once, when I am watching them, I see the barn owl again, white and shining in the gloom. It swoops to one of the big oaks at the top edge of the field on my right, where I know there is an old owl box, although we thought it abandoned long ago. When it reaches the tree there is a high-pitched screech of another owl from inside it and I realize it must have a mate.

Later, when the snow falls thick and deep, I worry it will starve the owl to death like it does so many other birds. The days bring frost, iced puddles and the valley draped with snow and thick fog. Then one day we are driving down the path to the road – *slowly, slowly* – when a great white shadow crosses the windscreen in front of me. 'Look!' I cry, and brake. The owl settles on the gate post right next to us and looks directly through my window. I try to stop the children yelling and jumping as they unstrap themselves and huddle around their window in the back. The owl doesn't seem to care. It stays there for what seems like a long time. I pretend to myself that it knew I had been worried for it. Then it takes off, up over the snow, which glows blue-white in the dusk, silhouetted against the red of the setting sun.

*It works*, I think. This – the animals grazing and bringing life with them, the long grasses at the field edges to hide voles and shrews for owls to eat, the new hedgerows – it actually works. I know that I am at the surface of discovering how and why. I also know we are probably getting some things wrong. We have planted the wrong shrub, or grazed sheep or cows in areas we should not, and not grazed them in areas we should. This feeling of muddling around in the half-light of knowledge makes me see how easy it is to think you're doing the right thing while causing harm. Many farmers thought they were making good choices, or at least that they were doing nothing wrong. Time has shown the opposite. Now it is the next generation that finds itself bearing the blame. I can see why some are so tired of this they've stopped listening altogether. But there were some farmers who could see what the future held long before others did. They bore the ridicule of difference, but now what they have learned might contain answers for those who mocked them. I have met a few farmers who chose to farm with nature rather than against it when everyone thought them crazy, but it is Sam and George who have taught me the most. They taught me to be suspicious of solutions that claim to have no consequences. They taught me that history has as many answers as the future. They taught me technical knowledge of the land, plants and animals but, perhaps more than this, they taught me what it is to know yourself, to keep your head, to trust your gut and your instinct and to always try to do what you think is right.

*There is a revolution afoot in farming. There's a lot of people waking up to the fact that there is quite a lot in farming that is very interesting: there's a combination of biology and chemistry and economics and you've got to get it all right for it to work . . . There's a lot of people who spend their working careers sitting in front of a screen and there's no fulfilment in that — working for corporates, there's no satisfaction. I've done it, I've been there. I realized: this is not a life for anybody . . . And to come back on the land and be here . . . The time is now, really.*

Andrew, farmer

# 10. Sam and George

The first thing Sam notices about the car is how clean it is. It gleams as it drives slowly up the winding pathway, pillar-box red, its colour so bright and out of place that it may as well be a spaceship arriving. Sam stands in the doorway of the house he shares with his brother. He is thirty-three, not yet married, and in charge of 120 acres of Suffolk land that his father says he is finally ready to hand over. He isn't, as it turns out, but that's another story.

The red car pulls into the driveway and parks alongside Sam's truck. The comparison is not a kind one. Two men – one the same age as Sam, one older – climb out. They are wearing the kind of uniform that makes them immediately recognizable as land agents. Mustard-coloured corduroy trousers. Checked shirt and tie. Olive green Barbour coat. Brown shoes, which they now swap for wellington boots from the boot of their car.

Sam waves and walks over to them.

'Morning, David.'

'Hello, Sam.' David leans forward and shakes his hand. 'Sam, this is my boss, Paul Carter.'

The older man leans forward to shake Sam's hand with a firm grip and a tight smile. It is the kind of smile that makes Sam realize Paul Carter might think this visit a bit of a waste of his time.

'Thanks for coming, Mr Carter. I appreciate it. David said you might give me some advice about what to do with the place.'

Paul Carter looks around the yard at the front of the house.

'Yes, well, I can certainly try.'

'Want a cup of tea before we go?' Sam gestures towards the front door.

'No, no,' says Paul Carter briskly, slamming shut the boot of his clean, red car. 'Best get on. Busy day.'

'Of course. Sorry. Thanks again for taking the trouble.'

'No problem. Right, then. Let's have a look round.'

It doesn't take long for the men to *have a look round* because Sam's farm – with its patchwork of fields laid down to crops and pasture for their flock of sheep – is small. This would not have been the case a few decades back. Once it would have been the average size for the kind of mixed farm that produced enough income to provide for several families and the community around it. The farm had been bought by Sam's father from Old Man Jones, who had lived and farmed there all his life. Old Man Jones had been a tenant of the Church and then, thanks to a good clover harvest, had managed to buy the land. The old farmer had kept going through all the changes the last century had brought, holding on to the farm when others around him were selling up. In the 1950s, when Old Man Jones was farming, there were just over 450,000 farms in the United Kingdom. Fifty years later, despite all the public money and all the wonder

260

chemicals and machinery, over half had gone out of business, a figure that would continue to rise. He watched small farmers ousted by those with the ability to levy what they owned to overpay for land, but kept his counsel and just carried on. Then Old Man Jones met Sam's father. Each liked the other, so when finally it came time for Old Man Jones to sell, he shunned the offers from big neighbouring landowners and sold his farm to Sam's father instead.

As Sam went to work in Australia, an agricultural revolution rolled across the countryside and farming empires swelled. By the time Sam came back to study agriculture at university there was a prophecy that, one day, thanks to this new combined gift of machinery and chemistry, farmers might actually be able to harvest four tonnes of wheat from an acre of land. Sam's father was proud of the one tonne per acre he coaxed from their difficult clay soil, but when chemistry overtook biology, the one tonne soon became two, then three. The four-tonne an acre fantasy, once on the same shelf as flying cars and robots, was achieved by a UK farm by the time Sam graduated.[1]

Afterwards, while he was waiting to take over the farm, Sam found his own part to play in this new farming revolution, working as a merchant selling seeds and agro-chemicals. Sam would arrive at farms and, from the thick catalogue under his arm, always be able to find something cheap and cheerful from a bag or bottle to solve every kind of farming problem. Everyone knew that often all it took was one farmer to buy and try before their neighbours would pass the field and wonder why theirs didn't look as good. Farming displays someone's work like almost no other occupation. Each visible field is an invitation for comment: good and bad. A farmer's ability, and so their worth, and so their self-worth, is weighed and measured by everyone and anyone who walks past their fields, whether they know what they are talking about, or not. When it comes to innovation or experimentation, only the brave or stupid will be the first to try. Few are willing to risk being the one whose farm everyone crowds around to point and laugh at.

Now Sam is going to take over the farm just as everything is getting bigger – machinery, animals, fields, barns, yields, ambitions. Everything except his farm. There is no adjoining land available to buy and, besides, any land that does come up for sale is bought by estate-holders or companies or pension funds that see it as an investment with a steady return of public money. So David, a friend from university, said he would help. He'd bring his boss to have a look, for although Paul Carter was more used to dealing with large estates and farms that had begun to call themselves 'agri-businesses', he'd no doubt have some ideas about what Sam could do.

Now, watching Paul Carter scrape mud from his boots, Sam is not so sure.

'I mean, I know it's probably all rather small compared to what you're used to . . .'

'Look,' Paul Carter sits on the lip of the car's boot, the lid open above him, and levers off his wellingtons. 'I appreciate you want to try your hand at this, but farms like this . . . Well, they just don't fit any more. They are just too small to work.'

He slips on his brown shoes, stands up and looks directly at Sam.

'I've seen a lot of farms and I've seen a lot of changes. All I can give you is my best advice. I'm sorry, Sam. This place . . .' He sweeps his hand around him. 'This won't work.'

He puts out his hand for Sam to shake it. Sam takes it, unsure what to say. Paul Carter fills the space for him. 'A beautiful part of the world here. Lovely, I grant you. But you know, times have changed. That's modern farming for you.'

David, standing next to him, leans forward, reaching to shake Sam's hand.

'Bye, Sam; thanks for showing us around.' He looks embarrassed, but Sam is unsure whether this is because of Paul Carter, or because of Sam's farm.

'Well, thanks for coming,' he says. 'Yes. I'll think on it. Thanks again.'

The two men drive off in the clean red car. Sam watches them go, then rather than going back into the house, he walks down the driveway and cuts right, through a path before the road. Beyond the path is the farmyard and, on the right-hand side, the old Dutch barn his father built. He looks at it and hears Paul Carter's words ring inside his head.

*This won't work.*

But what Paul Carter doesn't know is that Sam has two advantages. The first is being the kind of man Sam is. The second is that his friend, George, who has mostly been working abroad, is about to come home so the two men can work this farm together. Sam and George have known one another since university, longer than the women they will later marry, and will tease and bicker with a fondness played out in a thousand ways so small and familiar that they no longer really notice they are doing it. Their lives, and their family's lives, will become woven in and around one another's and this farm. And it will be these two combined – personality and partnership – that will prove Paul Carter wrong.

It is January, fifteen years later and the beginning of a new millennium when the piglets arrive, tumbling down the trailer's tailboard into the farmyard alongside the concrete barn where the old Dutch one had once stood. It was one of the first things George helped build when he came home and the two men set about planning how to make the farm work in an industry that seemed not to want it to do so. The farm had crops and a few sheep grazing the pasture, but Sam and George soon realized this alone wouldn't be enough to keep the farm going. So, as everyone else was becoming specialist, they decided to do the opposite. Getting into pigs seemed an obvious thing to do. There were once always pigs on farms around here, for this is pig country. The cottages within the winding villages of this part of East Anglia are still painted a crayon-like porcine pink. Locally it is said that the pigment for the paint once came from the blood of the pigs that lived in the fields around them. It was mixed into lime wash and painted onto the walls, a tradition going back as far as the fourteenth century. Each farm or smallholding would have had pigs to eat the scraps and provide rich manure for the fields, the farmers buying food each week from a wagon that drove around the country lanes delivering bags of feed labelled 'Hog's grub for sows an' littlens'.[2]

On the sandy soils of the county's coast pigs could live happily outside on free-draining land without destroying it. But the farm here is clay, heavy soil, easily churned to sticky mud by the animals' sharp-toed trotters and rootling snouts so, for the months that they are here, the pigs must be kept inside and fattened in the converted barn.

In the farmyard the latest batch of piglets run about, a pink confused mass of them, fast and nimble footed. Sam and George, now less fast and nimble footed, sweat and swear as they try to round them up. Finally, the men separate them into the pens inside the barn, where the weaners will live until they are fully grown.

The two men do not own these animals. Instead they are paid for their bed and breakfast by the company that do, and so they have all of the responsibility, but little of the autonomy. They must house and feed them and dose them with whatever antibiotics the company asks them to at the first sign of any illness, and sometimes beforehand, just in case.

Even though this is pig country, these pigs were not born here. They have come from Scotland, a journey of over five hundred miles. Now they will stay for just twenty weeks to be fattened by the food Sam and George bring them before another lorry comes to collect them. Then they will be driven hundreds of miles to Yorkshire to be slaughtered. There is a local abattoir half an hour away from the farm that could have done the job, but it's not the one that holds the contract. Pig business is big business. Each pig's carcass will then be sent to another factory for processing, then vacuum-packed into sausages and bacon and cuts that, mostly, look nothing like the animal that they have come from. Then they will be driven to a distribution centre before finally making their way onto supermarket shelves and the plates of people who haven't actually seen a pig in a very long time. From birth to bacon in six short months. Efficiency maximized. And yet also: not.

But, over the last decade, the pigs have been changing. Whereas once they were calm around the men, now they are edgy, jittery, dashing away and crashing into one another. Once they would have made a litter for themselves in the corner to keep their bed-straw clean. Now they just muck everywhere, making their cleaning out that much harder.

One hot spring day a few months after they arrive, and with the pigs now nearly full grown, George goes inside the shed to take out the dung and spread new straw. The air is thick with heat and dust and ammonia. Each man has a covering over his face to stop him gagging, although it is mostly George who does this work as Sam has farmers' lungs after decades of breathing in chaff and dust and fumes.

George bends and shoves his pitchfork into the straw and dust swirls up around him. As he does so, Sam's voice rings out, cutting through the gloom of the pig shed.

'Watch iiiiiiit!'

George hears Sam's cry but there's not enough time for him to move out of the way and the pig's flank sends him smacking into the concrete wall of the shed. The other pigs inside the pen, picking up on the switch in atmosphere, begin to scrabble about, running at the walls. George, the air knocked out of him, staggers to the stall door. When he is through he leans his tall frame against it, fork in one hand, the other on his ribs.

'All right?' Sam says, shutting the stall door behind him.

George nods and winces, trying to catch his breath.

'Madder and madder, these pigs are getting.'

'Must be the heat,' Sam says.

George nods, but both men know it's not the heat. It is the animals that are changing. It is not the first time it has happened. These new kinds of pig are now the ones bred for meat and confinement and to grow fat quick. George is not a superstitious man, but as he leans against the wall the whole business starts to feel rotten to him. You cannot, he thinks, breed for just one thing without losing another. There is a cost to every gain, if you're prepared to see it.

There is something else nagging at him. That in all honesty, this doesn't really feel like farming.

All George had ever wanted to be was a farmer. Whereas Sam's future had been planned out for him like so many other farmers', George had decided upon it. He loves farming. He loves the challenge of it, the freedom of it, the feel of it. Or rather, he did.

Afterwards, they will not remember who said it first, but there's little argument in the decision they then take which means that, when the lorry later comes to collect the pigs, it will not return to the farm again with new ones.

A week after the final lot of pigs have gone, the two men sit on a paved terrace in Sam's garden. The garden grass is fringed with small wild daffodils, whose bonnets nod when the breeze blows and remind the men of the Welsh university where they met and were taught how to become farmers, over two decades ago now, when the world seemed very different.

It is the turn of the millennium and the air feels full of change. To make a dent in all the food mountains, European policymakers have said they will start paying farmers according to how much land they own rather than how much food they make. Guaranteed prices will start to be cut. In a few months' time, a leading market report will put it coldly: 'Despite cutting costs and tightening their belts . . . farmers have suffered the lowest average incomes since our survey began eleven years ago.' Farmers are going to have to scale up further, consolidate, 'sweat their assets'.[3]

In this new century it is survival of the biggest. Small can't last any longer. The small will be eaten by the big and that's the law of the land, that's commercial forces, that's global markets, that's economies of scale. Call it what you will: it is unarguable and unstoppable.

Sam and George sit and watch the dusk fall. Beyond the hedges that run around the garden's perimeter are fields of still-green wheat and barley. In one of them a tractor is spraying crops, as the men had sprayed their own that week. Between the men is a wrought-iron garden table, and on it, two cold bottles of beer. Upstairs, two children sleep, as they do in George's house nearby, not knowing that their future is about to be caught up with the decision the men will make that evening.

Afterwards, it's hard to say what their tipping point is. Maybe it is because the new millennium has begun. Maybe it's because both men are in their forties and are now fathers who are feeling their responsibilities. Maybe it's news of fuel strikes and farmers' incomes crashing. Maybe it's because of what had happened with the pigs. Maybe it's the memory of Paul Carter in his shiny red car saying 'This won't work' and feeling like he may be about to be proved right. Or maybe it's all of this: a collection of clues that all end up pointing the same way.

No matter what the catalyst, it is George who says it first.

'I think . . . you know . . . I think it might be time for a change.'

What George means is: *We need to change, for the world is changed.* What George means is: *Is there really a future in this way of farming?* What George means is: *Is this even really proper farming?*

He is not the only one asking questions. The environmental cost of intensive farming has begun to be noticed. Farmland birds have declined more sharply

269

than birds in any other kind of habitat, their populations plummeting by over half in just thirty years.[4] Forty per cent of all insect species have declined since the 1960s.[5] Rivers are losing plant life and whole populations of freshwater fish because of algae blooms caused by nitrate runoff from fertilizers.[6] The reasons are complicated, because nature is complicated, and it is not just one factor that is responsible but many, all at once. The figures are frightening enough to make the public, and therefore the policymakers, want to see farming change. Now farmers are going to have to meet basic environmental legal standards and new environmental schemes are being introduced to try to help wildlife begin to recover.

Amidst all this, George had begun to read a little about organics. The government has launched an Organic Farming Scheme, and although the public funding is tiny compared with the rest of conventional agriculture, it has boosted a small emerging market, increasing every year. Demand is outstripping supply, so much so that, by the end of that year, 70 per cent of all organic food consumed in the UK will have to be imported.[7]

Organic farming has a loaded reputation, which makes it something of a joke to most other farmers, but to George it just seems interesting. To him it seems like knowing how to use the plants and earth and sun and rain, how to use the land – what's on it and in it, beneath it and above it – to understand how each affects the other, bound together as they are in cycles and rhythms, so as to produce food. There is something else, too. He can't see how their kind of farming – the one that has become 'conventional' – can last. George knows as a biological certainty that whatever is not killed will seed and breed and that plant will be more resistant than the last. It can't, thinks George, go on like this. There has got to be another way.

'I read that article. Did you see it? About going organic . . .'

Sam leans back in his chair and fixes his eyes on him. 'What? All that muck and magic?'

'Yeah.' George takes his beer from the table and looks away. 'All that. I reckon it might be interesting, though. Something different. Worth having a look. I read there's money in it.'

'Oh yeah, the grant,' says Sam. 'I heard about that.'

'Also, it'd be nice, wouldn't it? To just . . . do it ourselves. To farm how we wanted to. No one telling us what to do. No more bloody sales pitches. No more buy now, pay laters, because cor, don't you pay . . .'

Sam snorts. 'No one telling us what to plant and what to spray even though they've not sat in a tractor since Cirencester.'[8]

'Yeah, they can bugger off.'

Sam turns and stares into the garden. Tractors owned by the big farmers around them rumble a bass line into the dusk. He remembers Paul Carter's voice telling him *It won't work*. They cannot get bigger, he knows that. But maybe they can get different.

'Yeah,' he says. 'Yeah. Think you might be onto something.'

George waits.

'Yeah,' Sam says. 'All right, then. Why not? Let's try it.'

He reaches over and takes his beer from the table and lifts it towards George.

'Independence.'

George looks back and grins.

'Independence.'

Nearly a decade before the two men sat in their garden and changed their lives, a woman died. Her own farm was less than twenty miles away from theirs. Her name was Lady Eve Balfour and she was co-founder of the Soil Association, an organization that helped create a regulated market for organic food in the United Kingdom.

Born in 1898, Eve Balfour decided she wanted to be a farmer when she was twelve years old, at a time when the daughters of the nobility didn't really become farmers, especially if their uncle was the prime minister. No matter. When she was seventeen she became one of the first women to enrol for a diploma in agriculture at Reading University. At nineteen (but pretending to be twenty-five, so she would be eligible) she managed a requisitioned farm during the First World War. As peace was signed in 1919, she bought a farm with her sister with their intended inheritance. Two decades later, as peace was about to crumble and with the country on the brink of a second world war, she began the first long-term, side-by-side scientific comparison of organic- and chemical-based farming.

Balfour took soil samples every single month, from

every single field, for twenty-one years. She gave the soil samples to a biochemist to test them for a wide range of nutrients and bacteria. The results astonished her. The field with the highest humus content, and the longest history of being without chemicals or other inputs, had as much as ten times more available phosphate during the growing season than the dormant period. Potash and nitrogen followed the same pattern. There was, she concluded, a seasonal release of minerals that took place in biologically active soil. The soil, in other words, was changeable. It was alive.[9]

Balfour was one of a small group who, during the interwar years, questioned the long-term consequences of chemical farming. Other outliers were coming to the same conclusion. One was Sir Albert Howard, a botanist who moved from England to India in 1905 to teach Western agricultural techniques.[10] After arriving there he found his views challenged by traditional Indian farming practices. Indigenous people the world over had been farming land long before the arrival of European colonists and their methods of growing focused on natural cycles and stewarding healthy ecosystems. Accumulated over millennia, these techniques grew healthy, disease-resistant plants. It was Howard who found himself converted, rather than the other way round. He stayed in India for over two decades, learning from Indian farmers. In 1940 he published *An Agricultural Testament*, a book that spoke of soil as a living organism and which emphasized the importance of building up organic matter within it. Howard saw that the health of

soil humus was connected to the health of crops, livestock and the people who ate both. The system he
learned became known as 'organic' because it had a
complex, but necessary, interrelationship of parts, similar to a living organism.[11]

Eve Balfour's *The Living Soil* was published three years
after Howard's book, and contained an examination of
the results of her experiments with chemical and non-
chemical farming and soil testing. But then, in 1947, the
Agriculture Act was passed. Born of fear of loss of self-
sufficiency and nearly a decade of hunger, the Act
confirmed the government's commitment to creating a
mechanized, intensive farming system and food production. By the time Eve Balfour died four decades later, it
is hard to overestimate how deep the suspicion towards
organic agriculture ran. Balfour's book was long since
out of print. The so-called Green Revolution was saving
millions from starvation with chemistry and machinery.
Organics now was anti-science; a cult for cranks and
hippies. It could not, the majority maintained, feed a
growing world.

Not everyone thought Balfour should be dismissed.
Shortly before she died in 1990, a junior minister in what
was then the Department of Agriculture thought that all
Balfour had achieved over her lifetime made her deserv-
ing of an honour. He added her name to the list. So
badly was organic farming thought of that when her
name was discovered a senior civil servant complained
'No, no, Minister! We *cannot* have any muck and magic
here.' The junior minister insisted Balfour's name be

included. But when he double checked the final list he told me he found that the civil servant had unilaterally removed it anyway. He took the list to the prime minister's office and made his case. Eve Balfour was told by an official sent by that junior minster that she had been awarded an OBE very shortly before she died. She probably did not know about the battle fought within the corridors of power, but it would not have surprised her. She was used to being thought of as a crank. She never minded. After all, a crank, as Balfour once said, was a small and inexpensive piece of equipment that causes revolutions.

Ten years after Eve Balfour died, Sam and George set about relearning how to farm, unknowingly using methods she had tried less than twenty miles away, fifty years beforehand. They go into their attics and dig out notes from their old soil-science lecturer at university, so ancient that he died before they graduated and the only one who had taught them a biological, rather than a chemical, way of treating the soil. Now, years after graduation, they find themselves reading about medieval field rotations – where different crops were planted in strips so that if a plant was diseased or eaten its neighbour would stop the spread. They read about Thomas Coke, 'Coke of Norfolk', who cautioned that 'no two white straw crops [be sown] one after the other' to stop the soil from becoming worn out and to help prevent weeds and disease. For fertilizer they must use animal muck from their sheep, and grow plants like clovers and vetch beneath their crops, as these have nodules on their roots and take nitrogen from the air and fix it into the soil so the crop can use it to grow. The more they read, the more they remember. This is how it used to be done. Now, in the new millennium, they are relearning something old to make it new.

Their government is as sceptical of all this old-fashioned stuff as other farmers are. The year before Sam and George become organic farmers, Tony Blair's government spends £52 million on developing GM crops and £13 million on improving the profile of the biotech industry. It spends £1.7 million on promoting organic farming.[12]

It is hard at first to make it pay and they, like most other farmers, take on extra work farming other people's land. None of these landowners want their land to become organic, and so the men carry on spraying the chemicals they no longer use on their own farm. Without really intending to, they find that they are now running their own experiment, able to compare the crops they grow with chemicals to the ones they grow without them.

From each failure they learn something new. If they crop a field too much and reduce the soil's fertility then the weeds that thrive in this kind of soil begin to take over. Docks grow deep tap roots in patches of bare compressed soil – nature's way of subsoiling – and they try to pull them up, for they are taking up valuable grazing space, but soon the docks overtake them and they give in. Then one day they discover enough docks have grown to attract a colony of dock beetles, which have eaten the docks into wispy leaf skeletons and means they are no longer a problem. The clover they grow to provide the nitrogen their crops need is swamped by fat hen, but then they graze the sheep on it. At first it looks like they have eaten it to death but the cutting of the clover brings it back to life. Soon their fields are covered in small red-and-white tufts and the sound of bees buzzing around them.

They start experimenting with old wheats – heritage wheats – tall with long wide flag leaves, which look so different from the modern varieties in the fields around them whose height is controlled with genetics and sprayed growth suppressant.[13] They let the sheep graze the young wheat when it is green, shortening it so the sheep grow fat and the tall wheat doesn't fall over in the wind and rain. The sheep also eat off any disease, so when the crop grows back it does so clean and strong. They begin to realize that without artificial nitrogen boosting it, the old wheats grow much slower to start with, but by the time harvest comes they have caught up. This means that in the warm damp spring they still have thin stalks and an open canopy that allows the air to pass between them so they get fewer of the diseases other farmers spray off with fungicide. They find too that the roots of the old wheats are good at foraging for nutrients whereas the new wheats are bred for inputs and struggle to thrive without them. They get less off the field than others do but then their crop has grown for little and is sold for more. Still, they keep quiet when everyone is boasting of yields, even though they are the ones whose margins make a profit.

They learn that being an organic farmer surrounded by those who are not can be a problem. A newly emerging spring crop that looks well on the Friday is destroyed by the Monday after neighbouring fields are sprayed and the insects driven onto their unsprayed fields instead. Because they cannot routinely worm their sheep every six weeks, as some farmers do, they have to keep them off their old muck and break the cycle of the worm eggs laid in it, moving them onto fresh grass regularly. But they also learn that the greater the variety of plants they grow, the better they do alongside one another, especially when the weather tricks them with droughts, floods and gales.

They discover that farming this way means they must have animals, for the crops and fallow grass always do better with them, so after a time they add cows to the sheep – small, hardy ones that are good at looking after themselves and are kept as much for what they do to the land as for their meat.

Sam and George soon realize that while organic farming may give them *Independence!* it comes with more don't, can't, shouldn't rules than even their own government can dream up. The yearly inspections mean every seed label and grain store is checked to make sure there is no possibility of cross-contamination. There are higher standards of animal welfare to comply with. Organic cows, grown for either milk or meat, have to be able to graze outdoors for as long as the weather allows.

Chickens must have more space than is required by law. Pigs must be kept outdoors, and farrowing crates – where a sow is confined for birth and feeding her piglets – are banned. Medicines are only to be given when necessary, never preventively, and their withdrawal period doubled. If there is a pest or disease problem that stumps them, then they are able to use a non-synthetic product from a specified list, but only as a method of last resort and they have to record each one into the daily farming diary they must keep for their inspections.

But, despite all this, it's their choice what to grow and how to grow it. Their failures are entirely owned by them, as are their successes. It is, they discover, actually a pretty good feeling. Later it occurs to them that, no matter how hard the trial and error is, neither of them will ever mention giving up. For there's something the two men have learned to love. This kind of farming is interesting. Each year they feel like they learn something else, not just about the farm but about the way the world works. Leave it, they learn. Leave it. Be patient. Wait and see. They learn to rethink what a field should look like. They learn not to worry about the weeds, or worry about other farmers worrying about their weeds.

Still, they are left in no doubt that others think them mad. One friend, visiting from the dry sandy soils of vegetable-growing country in Essex, scoffs to Sam, 'What's growing in there – wheat or oats? Do you even know? Can you even *tell*?' Passing farmers laugh at the adapted machinery and tut about overgrown hedges, weeds, thistles, docks, worrying their own land will become infected. Many still look at the farm and think *this won't work*.

One summer, at the end of harvest, two local men lean on a gate watching Sam combine. The older one is dressed in a short-sleeved red-checked shirt stretched tight against his barrel stomach. He wears khaki shorts with side pockets that he puts on in March and does not take off until snow or frost makes him. The younger man is in his early twenties, red-headed and bespectacled. They are different in almost every way but the younger one loves the older like a brother. The older man takes the younger out catching rabbits with ferrets and with his hard-working hands, teaches him how to weave delicate stick fences to protect the bee and pyramid orchids they find in the fields. He teaches him which wood burns best and talks about the different smells each log gives as though he were a Michelin chef lifting a lid from a pot of fine sauce. He has taught him about both death and life and the countryside too, for both are found within it. Now the younger man looks up at the older one, grinning into his nut-brown face as he leans conspiratorially towards him.

'Lor' knows what he's harvesting there this year – a great crop o' thistles by the look of it.' He lets out a belt of laughter. The younger man laughs too, although not as loudly. He is shy about joining in because part of him is curious. He thinks it is quite brave to do something few others are, and to carry on even when they are mocking you. The younger man does not know then, of course, that he will marry a woman who will write a story about

Sam and George, and that she and he will decide to turn over fields to muck and magic themselves one day.

'Look at this, Sarah!' says Sam, looking at his phone, grinning. He laughs a lot, Sam. His face is fanned with the proof of it and an hour in his company is pure medicine. He has a smartphone and is therefore largely responsible for George's communication too. George has a mobile but it isn't especially smart, and he tends to handle it as though it were made of something explosive. Sometimes Sam teases him about being a Luddite, which does George a disservice, given it was his foresight that means the two men now find themselves at the forefront of this new farming revolution. I don't think either of them sees it this way but, nonetheless, they are the ones getting messages from progressive urbanites who have bought their wheat grains from a hip London bakery and planted the seeds in their city allotment in London.

Sam holds up the phone with the message from the progressive urbanite. 'She wants to visit the farm!' Sam howls with laughter. We are standing in his yard and I wonder what she will think of it when she comes. Does she expect something from *The Darling Buds of May*, or *Countryfile*, or a farm from the books we still read to our children that allow us to pretend our food comes from a place where a farmer and his wife (never the farmer) and their children live in a honeysuckle-strewn farmhouse. The farmers in those stories rear a few pigs, sheep, horses, chickens and ducks on little fields bordered by blowsy hedgerows and have a big red barn. They eat wholesomely, which accounts for their ruddy cheeks and jolly girths.

Sam and George's farmyard, however, looks like most farmyards I have been to. It has two barns – one with sides, one without – in which is collected the usual assortment of farm machinery. There is an old cabless 1960s Fordson Major tractor waiting for a few spare months which never come for someone to restore it. There is a pile of tyres, some cylinders of straw bound up with netting, spare metal hurdles on the ground to pen in stock and a trailer to transport the sheep to new grazing. One barn has a few stalls to house sick sheep or calves. The old pig barn is now a nursery for lambs, some of which have already arrived: pairs of lambs now sleep alongside their mothers, who stand and eyeball me and stamp their feet when I lean over the stall door. At the end of the barn is an old sofa, a sink and shelves full of ear tags and colostrum powder and elastic bands and instant coffee, where George will come and sleep on lambing nights. Sometimes there is a copy of *Farmers Weekly* knocking about with articles like 'Latest Hike in Nitrogen Prices' and 'Tips to Tackle the Rising Cost of Chemicals' and 'Webinar: How to Stay Profitable in the Face of Change'. It doesn't look especially bucolic or wholesome or like somewhere in a book. It just looks like a farm.

There are some differences, though. In the field opposite the yard are peas held up by enough weeds to make conventional farmers faint, but which George swears help lift the crop up off the floor and which form some sort of botanical partnership so that, when the crop is harvested, they find it is some of the best they've ever grown and get a premium price for it. None of their crops go for animal feed – all are high enough in quality for us to eat them. There are also machines you don't find on other farms. There is the weeder, an oversized multi-pronged hoe that teases out the weeds which grow between the crops, with thin metal spikes so sharp that I wince when the children run past it. And then, in the open-sided barn, is the experiment I've come to see.

When I look at the machine it doesn't look like the holy grail. It looks, with its row of bent arms, its bright red hopper and waving tubes, like something from *Star Trek*. But it is. If it works, this machine will allow a crop to be planted directly into the ground without ploughing the earth, and grown without any chemicals or artificial fertilizers.

If they were to buy such a machine it would cost them hundreds of thousands of pounds. So George has used his skill at engineering to do what he has always done: adapt something for his purpose. He has taken a drill he bought for a fraction of the money and fitted other parts to it himself using, amongst other things, an old motor from a car's windscreen wipers. His invention probably won't work straight away. They will have to experiment and try again. But they will follow the rule the pair have set themselves. Don't worry about what others are doing; just do what you think is right.

Sam and George look a little uneasy when I tell them I think they are at the forefront of a farming revolution.

Maybe they think I'm trying to flatter them and will tease one another about it – 'All right there, Nostradamus' – when I'm not there. This is the reason I say it. It is easy – or easier – to be different and take a risk if you have inherited a large estate. It is easy – or easier – to risk failure if the only way you will feed your family does not depend on it working.

Sam and George are not the only ones doing this. All over the country there are small farmers, whom few will ever know about, who kept going when others told them 'This won't work'. They don't have blogs or social media accounts or a logo. They just understood that the world was webbed, each part connected, and that breaking enough strands of silk would mean that eventually the whole structure would collapse. Some of those who realized this were small people with small voices, and they just hunkered down and accepted being called odd or old-fashioned. But some were big people with big voices, and even they were called anti-science, crazy, out of touch. In 2021 Prince Charles, who has spent decades being ridiculed for his predictions that the overuse of plastics and chemicals would end badly, received the *Farmers Weekly* Lifetime Achievement Award. The first line of the two-page article he wrote in response said he was 'surprised and delighted'. I imagine there may have been other words he wanted to use too.

Now, as government payments switch in a few short years from money for land to 'public money for public goods', as nitrogen prices treble and diesel prices soar, as chemicals are restricted and weeds and disease become resistant to the ones that are left, it is to the farmers who were once mocked that others find themselves turning for some of the answers to farming's future.

Unexpectedly, I find something less obvious in the choices Sam and George have made. In taking this path, they created a community. Loneliness is not exclusively a countryside affliction. It is as easy to feel it when surrounded by hundreds of people as by none (although harder – practically speaking – to cure). But Sam and George's taste for experimentation, the way they have made their farm alive with birds, flowers and insects that the people in their community enjoy, and the fact they mostly sell to people they know rather than into a global pile, means they have collected people around them and, in doing so, connected these people to their land. And, of course, they have each other. In an industry where loneliness and depression feed directly into the statistic that puts its suicide rate at double the national average, I suspect this counts for a lot.

Listening to them, I can see too how this kind of farming changes the way you think. It may not make you rich, it may not make you popular but it asks you to look for the truth. When you cannot reach for a bag or bottle to fix a problem you have to learn about its true cause. You must discover why something is happening, rather than try to treat only the symptoms and hope you might find a cure. When the rest of us are making decisions – about our mental health, or physical heath, or politics, or relationships, or society, or ourselves – this is probably a pretty good philosophy to learn to live by.

Once, George talks to me of reading Darwin. He says that Darwin saw the world as a living organism: it will always respond to anything you do and it will always even it up in the end. He is talking about farming but, sitting opposite him in the garden, I think instead of politics. I think of watching the pendulum swing to both left and right, making the world dizzy with rage, and wondering whether this means it will soon end up in the middle again, where it will stay long enough for people to forget and start the pendulum swinging again.

That I start to see the pattern of human behaviour echoed in the language of farming is maybe not so surprising. The Latin word for earth and soil – *humus* – and the Latin word for human – *humanus* – may come from the same root: a 6,000-year-old language called Proto-Indo-European. Its words 'dʰéǵʰōm' and 'dhéǵmō' mean 'earth, earthling, earthly being', as opposed to godly ones. It may be where the Latin *homo*, *humanus* and *humus* all have their origin, as well as the Latin *humilis* – humility – meaning someone who is lowly, humble – literally 'on the ground'.

One day I look up a passage from Genesis, the one often read at funerals, and find out that it says so much more than the 'from dust to dust' I thought it did.

> By the sweat of your face you shall eat bread
> until you return to the ground,
> for out of it you were taken;
> you are dust, and to dust you shall return.

We are designed to be earthed. We grow the food that keeps us alive from it. We will end up part of it. It's written into our languages. It's written into our prayer books. When I think about my old life and my new one I wonder whether it was our distance from the earth that brought about the hubris that caused its end. If this is true, there may be no better cure – no surer way of refinding humility – than learning how to farm, and few people better able to teach me than the ones who worked this all out long ago.

*I believe that these old ways, these are going to create food for us in this world we live in, where we have disease and the weather changing. This is more important than we realize.*

Tony, farmer

# 11. Year Five – Paying Attention

I drive up the path to the crop fields and realize I am nervous. I want this meeting with Richard to go well more than I had understood. In my phone notes I have written a list of questions to ask him. I want to learn from him, and with him, but I also know that when a sentence starts with 'why' it can sometimes sound less like a question and more like an accusation.

When I turn the bend at the top of the fields I see Richard is already parked up. It is about as bleak a January day as it is possible to get. A bitter wind whips over the fields so forcefully I have to push my car door hard to open it. We get out at the same time and walk towards one another as the grey sky starts to spit rain down on us. I grin at him as broadly as I can as the wind smacks my hair around my face. He grins back.

'Hi!'

'Hi!'

We stand in the wind wearing smiles and trying to read each other. It strikes me that maybe he too drove up the path with reservations. That he too wants this to be easy and conflict free, but that he also resents having to explain his decisions and choices after farming for more years than I have been alive. I wonder whether he already knows what my first question will be.

*Why have you ploughed the field?*

A few decades ago this question would have been absurd. A ploughed field is as symbolically tied to farming as a tractor or a cow. It is an image of our relationship with the earth: a black-and-white woodcut on the cover of an old book.

When I was a child I would balance on the icy top of a ploughed trench in winter, waiting to see how long it would take for the soil to give way and my boot to collapse into the earth. In my memory, the soil smells richer than perfume. I remember wrong. Cold earth smells of almost nothing. We all like to romanticize the land, it seems, even to ourselves.

Now, I see something different. I don't see rich brown earth turned up and readied for new seeds to be planted to grow within it. Instead I imagine the carbon that had been locked into the soil being released into the atmosphere, gusting up like those thermal images used to frighten people about heat escaping from uninsulated homes. I imagine the sliced-up worms, broken soil structure and disrupted fungal pathways. I imagine rain pooling within the peaked soil trenches, washing to the edge of the field and into the ditches, taking topsoil and nutrients with it. Now when I see a ploughed field it just looks exposed, almost shocking, like a great brown wound.

The day before, I had checked through the form we had filled out to apply for our public stewardship funding. In it we had promised those giving us money that within these fields we would implement 'organic, min-till and regenerative farm practices across the farm'. 'Regenerative' still means different things to different people, but most agree it stands on five main pillars: growing a variety and diversity of crops rather than a monoculture, using grazing livestock, keeping the soil covered as much as possible, minimizing soil disturbance and trying to ensure there are living roots in the ground all year round.[1] These last three mean a farmer cannot plough.

The plants that were growing in this field before it was ploughed were the cover crop.[2] They were only ever intended to be temporary. But before the crop that earns the money can be planted, the cover crop has to go. For most farmers this is easy: they can kill the cover crop with glyphosate and then there is no need to plough it, as the seeds for the crop can be sown directly into the soil without turning the earth at all.

But we can't use glyphosate. We've committed to doing without it. Thirty-three countries have already banned the chemical – including our neighbours France and Germany – as lawsuits rage in America brought by those who claim it has given them cancer. Other farmers tell me they wouldn't be allowed to use it if it wasn't safe, but I read science which warns that the chemical finds its way through the soil into our water streams,[3] and is absorbed through the roots of other plants, inhibiting their ability to grow and weakening their ability to fight off pests and disease, meaning insecticides and fungicides have to be used to do this for them.[4] I read how, as weeds have become resistant to it, fifteen times more glyphosate is now used globally than when it was introduced nearly fifty years ago. My father tells me a story about arranging for a lime tree to be felled and its stump doused with glyphosate to stop it sprouting back. A year later the previously healthy tree growing alongside it died too. The chemical, passed through the trees' roots, had killed them both off.

Ben and I have convinced ourselves that farmers have grown food without glyphosate for many thousands of years; that farmers around the world do without it now. But then, says Richard, you have to plough. You plough to kill the cover crop then turn it into the earth where it rots into a green manure that feeds the next crop. You plough to kill the weeds and bury any seeds these weeds have dropped to make sure they do not grow. And you plough to whip the earth into peaks that the winter frost will touch with icy magic, expanding and contracting the water molecules within the soil particles so that it crumbles, ready for planting. You plough for the same reason farmers have ploughed for millennia: to reset the earth before a new harvest can begin.

I did try to find another way. Could we graze the cover crop off with sheep? Could we break the cover crop with a crimp roller – a machine that looks like a giant's metal spiked rolling pin – which kills it and turns it into a dead mulch? Could we mow the cover crop, leaving the remains to rot down into the earth, and hope that the cold frost would kill the rest? I had drawn two columns on a sheet of paper – no-till non-organic farming versus organic farming with ploughing. Neither wins. Articles and social media posts about farming make clear that there should be a winner. They are clear that there is a right way to farm, and a wrong way to farm. I am beginning to see it's a little more complicated than that.

Richard watches me glance over his shoulder at the ploughed field. He knows my question before I've asked it. 'Rain was coming,' he says in answer. They wouldn't have been able to get into the field once it had rained without damaging the land. The purple-flowered phacelia needed to be dealt with before it dropped its seeds and it was too late to organize sheep to graze it off. He'd asked another farmer but he'd promised his animals elsewhere. There's no such thing as farming on a schedule. You have to make the plan day by day as the weather lets you. Rain was coming. He'd had to make a decision. He had gone ahead and ploughed it.

He gestures towards the field with his hand. He is pleased with how it has gone: he'd got in before the weather turned and managed not to make a mess of the land. The few hard frosts have killed any last weeds off and gifted us a good seed bed. All in all, he says, it is a job well done.

I don't know enough to be sure whether I should agree with him or not. But later that year when oats grow up in that field, straight and tall without being smothered by weeds or wiped out by disease, Richard is proved right.

On a map, the field edges are small slivers. It is so easy — too easy — to take a red pen and draw the line.

*We'll take this bit out of production here. Square off that corner. Take out that chunk.*

Our crops now will only be planted in rectangles. All the fields' awkward corners will be used to grow plants and flowers that will, we hope, bring predatory insects to eat any aphids that might eat our crops, now we cannot spray them. Today the pen marks are becoming boundaries. We are walking the fields in the February rain with a collection of stakes under one arm and a mallet under the other. Except now the land we are taking out of production doesn't look like slivers. Now it looks huge.

It rained all day yesterday. The soil sticks mud-heavy to the soles of my boots. Each foot comes away from the earth with a *puck*. It feels like I'm tied to weights. I stumble to the field edge and take the photocopied scrap of map from my pocket. Ben is in the middle, but the wind is too strong for me to hear his shouts. We wave and point to communicate how far into the field the new boundary lies.

*Further . . .*

*Further . . .*

*MUCH further. Keep walking!*

*Bit more . . .*

*Stop! There.*

When we casually drew our pen across the map it seemed we were taking out small corners. Now I am standing in a great slab of land, land that could grow food and where we will now grow flowers. I hear the voices of a hundred old farmers ringing in my head: *Well, where's the sense in that?*

Ben bangs a stake into the earth. Each has been tipped in white paint, stark against the brown. We walk on and on and do the same in the corners of other fields. I think about Richard warning that the time and diesel it will take to drill the seeds may end up costing more than the stewardship payments. And, oh, the fiddle, he had said, like a line from one of the children's books. *What a faff and a fiddle.* I stand at the corner of the final field and hammer the last stake into the claggy earth, close my eyes against the winter wind and try to remember why we are doing this.

I picture a Cambridgeshire farmer called Martin Lines, whom I met nearly a year beforehand. I imagine him standing at the edge of his field after harvest, staring at a tractor ploughing up the earth. He has a voice ringing in his head too: this time, it's his dad's, *What are you doing there, standing around? Go on, go on; get going with something.* For his father's generation, standing around *not farming* was considered close to a sin and mostly curable with some recreational ploughing. The muscles in Martin's legs twitch but he grounds his feet to the earth. *I need to watch, Dad. I need to stop, and watch a bit.* The negotiation works. His father's voice quietens. The sound of the traffic on the road that divides his family farm from the large contracted one opposite competes against the noise of the machines. That contracted farm too is ploughing, although their field is so huge they have two tractors working in it at the same time. Martin knows they are being driven by seasonal contractors. They will be driving the tractors because they have a calendar that says they must. The dates have nothing to do with the weather. They have nothing to do with the soil. He has seen them combining wheat in the rain before.

He frowns into the autumn light. Inside the cab, a young lad is driving. The lad is over from Australia. Martin has hired him to plough this field but for some reason the boy seems unable to plough a straight furrow. Martin looks at the earth in the field. It's not too wet; not too dry. *What is he doing in there?* Martin can see no reason why the tractor is weaving all over the place. He curses under his breath. They used to always get old Jake, the farmhand, to plough this field. Old Jake knew how to get a furrow ramrod straight just by lining the tractor up with a point on the horizon, but then old Jake died and took this skill with him, and now they had this kid for the summer, used to the colossal machines of prairie land with so much tech he could just put his feet up and open a beer.

The tractor turns at the headland and makes its way back to where Martin is standing. He waves it down. The boy stops and leans across to open the door.

'Hey there, mate, everything all right?'

Other farmers would have torn strips off him. Martin hasn't got it in him.

'Go real slow on the next one will you?'

'Yeah, sure thing. No worries.'

What Martin does next is on instinct. Had he stopped to really think about it, he might have thought it a 'farm safety' example of how to get yourself killed. But he doesn't really think about it. As the tractor and plough turn slowly past him he runs, jumps and sits on the back of the plough, gripping on to the steel beneath him with both hands. From this position he can now see the boy through the back window of the tractor. He watches the boy's hands swing the wheel back and forth, jerking the tractor from left to right. Martin feels his frustration rise. Is this the future of farming now, he thinks; lads who can't even hold a tractor wheel straight?

He sits on the back of the plough, meaning to jump off when it reaches the headland. As he sits there he looks down and watches the steel blades run through the stubble, turning the earth into a rich brown ridge. It is then that he sees it. As the soil is scooped and lifted he sees last year's tractor tyre marks imprinted on the earth below. The pattern is as strong and clear as if it had just been made. Small pale strands of straw lie in the grooves: the stubble from last year's harvest. This stubble should have decomposed by now: broken down by microbes and worms. The tyre marks should have dissolved into the soil. The truth of what he is seeing hits him with a great force. This earth – his earth – is so empty of life it cannot even decompose plant matter any more. It has become so compressed, so biologically depleted, that the imprint of last year's harvest is preserved within it as clearly as if it were a fossil.

When the tractor slows at the headland Martin slips off the plough in a daze. He walks to the farm office and switches on the computer and stares at the balance sheet he had been looking at that morning. The input column – all that he buys in bags and bottles to make his crops grow – is larger than it was last year, and the year before that, and the year before that. *All this money*, he thinks. *We spend all this money, on all this stuff, and look at what it does.*

Cereal prices down; input prices up. It can only end one way. The farm would be lost, and lost on his watch. He sinks back in his desk chair, stares at the screen and wonders what to do.

I now picture myself sitting at a table in Martin's small office. Outside, the sunlight is streaming into his farmyard. I have my phone and a small notebook and pen and will later wish I had ignored the latter one and used the former to record, for Martin will talk so fast and say so many interesting things it will be impossible to write down or remember it all.

Martin sits on my right, pointing to the end of the room at a projector. It shows a map lit up in red and green and yellow, the patches of colour blurring at the edges. I fight an urge to turn away: maps, numbers, statistics. These are not my language. But I need to see it, to understand it, because Martin is telling me something extraordinary.

Statistics flash up on the screen and I scrabble to write them down in my notebook. He tells me how he knew his farm could not financially afford to keep going the way it was. He needed to change but he also needed a plan. So he started collecting data: loads of it.

He points to the areas of red and yellow on the map of his farm. The colours show how the soil changes from clay to sand and back again, even within just one field. He overlaid this soil data with information from the brain of his combine harvester and sprayer, able to record exactly how much crop was cut from every part of the field and what he had used to make it grow. He worked out how much it cost him, in time and diesel and seed and spray and fertilizer, to farm each square of the gridded fields. And then he took a pen and crossed out all the squares that earned him less than the money he spent on them.

The field with the wonky ploughing was never ploughed again. When I first see it, its soil has not been turned for ten years. Instead, Martin now grows a cover crop after harvest and over winter that is filled with plants like buckwheat, phacelia, sunflower, vetch, linseed, fodder radish, crimson clover. Then he sprays it off and drills the new crop directly into the ground.

Later we walk into a different field and he sweeps his arm towards it. A third of this field, he tells me, has been taken out of production. Much of it was shaded by a nearby wood so the crop never did well. He had already planted a six-metre grass margin to prevent chemicals and soil leaching into the watercourse but now, beyond that, is a massive chunk of field, nearly three acres, all given over to plants which some call weeds and others call wildflowers. At the other end of the field he is paid to plant a bird-seed mix to provide food for birds over winter. Squaring the field up of awkward corners has stopped any overlaps of inputs, Martin says. 'But here's the thing . . .' He turns to me, putting his hands in his jeans pockets. 'It makes me more money now than it did when I was cultivating the lot.'

He smiles. 'A funny thing happened in this field,' he says. 'I came out here on a Friday in early spring to check a crop of beans. Found aphids everywhere. I called my agronomist, who said I needed to spray an insecticide, otherwise black-bean aphids would have the whole crop. But it was wet and the wind was too high, it would have blown everywhere. So I was going to do it the next day, but I was away, and then the week after was rotten weather too, and somehow it all just got too busy. So ten days later, when it's finally dry, I get it all ready: load the sprayer up, drive it down here. Just before I started spraying I think, *I'd better have one more look*. So I get out of the tractor and check the field. But there aren't any aphids. They have all gone . . .'

'Gone?' I say, not understanding.

'Gone. Been eaten. This . . .' he nods his head at the triangle of land given over to wildflowers '. . . it was alive. Full of ladybirds, hoverflies, lacewings. Most of the things around here I don't even know the name of, to be honest with you. They'd had a feast on the aphids.' He laughs. 'So I just turned around and just drove everything back to the yard.'

What Martin learns makes him rethink everything. He discovers that not using insecticides makes no difference to his yield, although it does to his bank balance. He changes what he grows. Whereas once he planted wheat, oil seed rape, wheat, beans, all sown in winter, now he has no fixed rotation but tailors the crop to suit the ground that year, depending on weeds and soil health and weather. He grows a mix of wheat varieties to try to outwit pests and diseases. He grows clover underneath his beans; phacelia between the rows of his oil seed rape. Then Martin thinks, why just grow wildflowers at the edges? Why not grow them running through the fields? He shows me a field where every fourth tramline is green with the beginnings of flower strips. Later he will tell me that the yields next to his wildflower strips are between 5 and 20 per cent higher than in the middle of the field.

He tells me of the compromises. He is not organic. He still sprays. He sprays so he doesn't plough because, for now, that is the only way he can destroy the cover crop. Few can do this perfectly. Everyone must make an accommodation with reality.

That day, on the way back to the farm office, Martin and I walk through a small meadow adjacent to the unploughed field. As we step into the long grass it seems to rise up around us. A rippling cloud of butterflies and moths and insects fly into the air until we are surrounded by them. Martin grins. 'A special place, this is. I get loads of experts coming along to look at it. They've found all sorts in here. Things I wouldn't ever have spotted. I mean, we were never taught them, were we? I couldn't tell one bird from another – they were all just "small brown jobs" to me. But I'm beginning to learn how to spot them now. Once you know what you're looking at . . .' He trails off, moving forward through the cloud of tiny living things. 'Well, it just changes everything, really.'

Back in our field, hammering a stake into the ground, I realize that I now know exactly what he means.

I am with Wilfred at the bottom of the lower meadow by the river when I see the first swallow arrive. We are looking at the new hedges, bare-sticked after winter. The self-seeded oak trees, grown up in the meadow from acorns planted by rose-coloured jays, have been replanted in the gaps. I am trying to imagine what this field will look like when I am dead and the oaks have grown as tall as the one at the top of the field edge. Then I notice the swallows – one, then two, then three – appear over the hedge line. I whoop and shout that they have just arrived all the way from Africa, a journey of over 6,000 miles. Wilfred chases the birds as they dart about above us, waving at the sky – 'Hello! Hello!' – like he is at the arrivals terminal of an airport.

The spring arrives in technicolour after the hardest of winters. The song thrush that has sung outside my bed-room window, each verse so different from the last that it is hard to believe so plain a bird can create music so magical, is now joined each morning by others until they are loud enough to count as a chorus. Small purple cro-cuses push their way up through brown mud and daffodils fan themselves around the bottom of trees. They are not the only plants growing. So are the weeds. When we drive to meet Richard up at the crop fields I see him striding through the mottled green covering the fields and my stomach drops.

I am wrong to worry. Richard is buoyant. He reaches down and pulls one of the weeds up easily, showing us how shallow-rooted they are. They won't be a problem, he says, a tine weeder will lift them out.[5] 'Great,' I say, remembering how the threat of these weeds once haunted him and trying not to look as glad as I feel.

Within the weeds were oil seed rape plants grown up from seeds left behind after last year's crop. All have now been eaten by pigeons, leaving only the woody stalks behind. These plants and the weeds seem to have protected the soil. After the wettest of autumns we can walk across the field without any earth clagging onto the bottom of our boots.

A few weeks later the weeds are out and oats and barley drilled into the prepared soil. Richard is pleased with how the crops go in. 'Funny thing,' he says as we walk the field's boundary. 'I think those weeds actually kept the moisture in the top of the soil. It's what we want now, really, a good bit of moisture to get it all going.'

Weeds being useful. Now there's a thing.

Over spring the verges of the crop fields become banked with cow parsley, ox-eye daisies, yellow buttercups and the tiny white flowers of speedwell and stitchwort, which look like a fairy's garland. Lumps of chamomile grow up by the top of the big field. I pick great clumps of it and put it in a vase in our bedroom where it lasts for weeks and make the room smell of sweet hay. One day Ben talks to the people who live in the house opposite Mr Pink, at the bottom of the path. They tell him, in passing, how glad they are to see wildflowers growing up the banks of the path again. When they first moved here, a long time ago, flowers always used to grow along the path. It has been so long since they've seen them they had forgotten they ever did so.

I find a photograph I took last summer when we were

planning changes. At the time I had not really noticed the yellowing of the grass on the margin where it had been killed off with glyphosate or that there were no wildflowers growing underneath the oaks on the path opposite, where the wind must have carried the herbicide across. I look at the photo and then at the new ones I have taken and wonder how I could have missed it. But I did. It is, I now see, easy to accept a landscape as normal without realizing both how abundant and alive it once was, and that it can be again.

'It's here, just here . . .' Richard strides through the young barley towards a wooden paint-tipped marker, then bends in half and pushes the crop aside with his hands. I lean forward, trying to see. At the base of the crop I can see small round leaves of green clover.

'Just a little trial,' Richard says casually. 'Thought it might be interesting.'

He tells us that when he drilled the barley into the field he also sowed a test patch of clovers. This 'under sowing' can help prevent weeds and provide the crop with additional nitrogen, taken from the atmosphere and fixed into the soil by the clovers' roots. I stand upright, grinning, trying not to look surprised. But then, I shouldn't be. It seems this year might not just be changing our fields. It might be changing Richard, too.

As I stand there I think of another farmer I met, Nick Padwick: a farm manager on a large estate of just under 4,000 acres in north Norfolk. After a lifetime of farming one way – and being celebrated for it – he too has now changed everything, not just about the way he farms but also about the way he thinks.

On the day I meet Nick I struggle to keep up with his speed, both physical and mental. He strides into the middle of a field as I jog behind him and over to a metal contraption the size of a barrel. Nick tells me it is a weather station. It uses weather data and information from a probe running deep into the soil below it to assess the likelihood of disease in the crop. If it reaches a high chance of infection Nick will spray a fungicide before the disease can take hold, meaning they use far less spray. That year the farm will end up spending £38.37 per hectare on herbicides and fungicides. The year before they spent £226 per hectare.

Just over ten years earlier, Nick won an award for Farmer of the Year. At the time his farm was one of the biggest arable suppliers in the country and used all the modern chemicals and machinery available to grow oil seed rape and wheat in a block rotation. He was good at it, one of the best in the country. Now, though, he sees it differently. 'Forty harvests wasted,' he tells me. 'I've got just ten or so left to make up for it all.' I tell him he's crazy to think like this, but I'm not sure he believes me. He feels the weight of the past in a way few others do. For him, this new/old way of farming cannot happen fast enough.

Some from his old life criticize him when he writes about how much money this way of farming saves as though he is telling tales, but what Nick has understood is that this is not just about luring birds and bees and creatures back. This way of farming is also cheaper; much cheaper. His fixed costs have fallen by 40 per cent in just one year, and even if some don't want to believe him, the proof is all there in his balance sheet.[6] He seems defiant in the face of all the muttering but also, underneath the toughness, maybe a little bruised. It's not always easy to be the outcast after you've been in the centre.

Later, we stand together at the brow of a hill that loops down towards a road. Nick has been showing me how he is growing oats and peas in rows alongside one another. The nodules on the roots of the peas fix nitrogen from the atmosphere into the soil for the oats to use as they grow. Last year they used only a fraction of the artificial fertilizer they once would have done. He tells me something I've heard from other farmers: some strange alchemy happens when the plants are grown side by side which means they ripen together whereas, apart, they do not. This makes them able to be harvested at the same time and the peas sieved from the oats. The following year Nick will attempt something most farmers consider impossible: he will sow wheat into a pasture field without ploughing, and grow the wheat without spraying any pesticides or adding artificial nitrogen to help it grow.

That afternoon I stand and watch a gang of hares, maybe six or seven of them, race along the top of the field near a hedgerow. I think of the story Nick just told me of a hare that once ran under his sprayer boom before sitting at the edge of the field cleaning itself of the chemicals it was covered in, and the feeling in his gut when it did this. Within the hedge a chiff-chaff belts out its see-saw song. Nick looks down towards the road and I follow his gaze.

'We planted wildflower strips along that field edge. Last spring so many poppies came up that people driving along the road started slowing down to look.' He grins. 'I thought there'd be a crash. Their tyres left skid marks all the way along the verge. Everyone was saying, "What *are* they doing in there?" And I thought – *That's it! That's what I want to do here.* I want to make people stop and look. I want to make them really see. Then they will know it's possible to do this differently.'

Back in our own fields, I stand and look at Richard's new field trial and think how, so often, change has to be seen to be believed. But then again, change is also infectious. It is catching. All it takes is someone to be brave enough to start and soon the ideas will roll out to their neighbours, and their neighbours, and their neighbours, onwards in waves until there is no one left able to say 'that won't work here', because it already has.

All summer it rains and rains and rains. Around the country the moods of farmers darken with the clouds. Combines sit unused in sheds as their tin roofs rattle. We stand overlooking the barley, their stalks drooping down in a way they weren't a few weeks earlier. Now the day has finally dried up and the sun is out. I can hear a sound, hundreds and hundreds of sharp cracks, like a field full of tiny fireworks all going off at once. Richard tells me it is a good noise. It is the sound of the grains drying.

He tears back and forth from field to field. He wants to get on with it – just GET ON WITH IT! – but the moisture level of the crops is too high. If they are cut now the grains might rot in the shed and the harvest will be wasted.

We message him back and forth – *Today? Tomorrow?* Finally, finally, it is *'Now!'* I have learned a lot about farming in the last few years but, more than anything else, I have understood that farming is like childbirth. You can go ahead and make a plan, but be sure that nature will come along and ruin it. Still, in the end, when all is finished, you realize you have grown life from a seed and somehow you forget the pain. This contradiction makes farmers some of the most stressed and the most philosophical people I know.

We sit alongside Richard in the combine and watch it cut through the crops. I hadn't expected it to feel so good. These crops were grown with nothing but sun and rain. I had almost begun to believe that such a thing was mythical. Afterwards I will find that reaction in other farmers, whom I suspect think I am lying – that we must have used *something* – because they too have come to believe it is impossible. But here is the proof, pouring out of a funnel in a thick golden stream.

The sheep look like teddy bears. They are New Zealand Romneys, whose fluffy heads and beady eyes make them look like a child's toy. They are not our sheep, but they are on our land. The sheep will cut through the grasses, clover and herbs just as well as a machine but with no diesel, no soil compression, much more muck and at no cost. We will earn a small sum of money from the shepherd who owns them, although what the sheep will do for the land will be worth far more. I stand and watch them from behind the electric wire, shoulder deep in red clover flowers. It strikes me that this is probably the first time this land has had animals on it since the 1960s. Then I realize I am in the same corner that Ben and I went worm hunting last summer. Somehow it doesn't just look like a different field. It looks like an entirely different farm.

The sheep bounce away from me as I move around them. I walk up the slope and across the new margin, pass the line we marked out last autumn where a hedge will be planted. Skylarks, playing at pheasants, take off from the barley stubble seconds before I reach their ground nests, singing loudly in indignation. When I reach the top of the field where the pathway bends, Richard bounds over, beckoning me to look at some seed bags. We are planting two more cover-crop trials. He will drill mustard, clover and phacelia in one and mustard, phacelia, vetch and buckwheat in the other. The farm is now divided into experiments. Some plants have been drilled directly into the soil. Some will be sown with only the top part of the soil being turned. Some soil has been ploughed. And when the grazed cover crop has been ploughed up a portion will be left and winter wheat drilled directly into it to see if we can grow it without ploughing or chemicals in a technique called pasture cropping. As I said, change is contagious.

I leave Richard talking with Ben and go to stand under the oak tree at the top of the path where Mr Pink's camping chair sits. Now it overlooks the plants we have grown to feed birds over winter and the land is filled with flowers and insects. Aubrey, who has been climbing up and down the tractor's steps, comes over to stand by me. I take his hand and we wade into the middle of the plants as the air comes alive around us. We crouch down and try to see how many different types of caterpillars we can find and work out what butterflies they will become by looking at pictures on my phone. I show him how the old hedgerow alongside the field is still covered in red haws and berries because, for the first year in many, it hasn't been cut. We are about to tip into winter but somehow it doesn't feel bleak. It feels like there is colour and life everywhere.

Aubrey leaves me, slow-running back with a caterpillar he has balanced on a leaf to show his brother. I go back to the oak tree and lean against it. The leaves above me are starting to change colour, the connection between tree and leaf stems breaking and the leaves starved of water and food so that the tree can preserve its energy for winter, ready to grow new ones when the spring sun begins to shine again. The wind blows them around me, freckling the grass with yellow and brown. From where I am I can, if I turn half a circle, see all of the crop fields. I realize suddenly how it is not just that they look different. This whole place feels different. Once these fields around me just looked like land. Each one was a block of crop in shades of yellow planted into brown. It had hedgerows only at its boundary, margins of long grasses and a handful of old oaks but otherwise each field looked more or less the same. Now every part of it looks different. There are six fields where there were once three, each separated from their neighbours by thick green margins that are soon to be home to new hedgerows. In every corner there is a mixture of flowers, creatures in and on the land, below and above it, big and small. There is colour everywhere but there is also texture. Then I realize why it feels so changed. This is no longer just land. This is a farm.

Back in the pasture fields the two-year-old hedgerows now have oaks which are nearly as tall as me and I tell them so – 'Haven't you grown!' – as they wave back at me with autumn leaves.

The gaps within the hedgerows have been filled by thistles. I have watched them turn from purple into white silk fluff and into the brown of seed head, which I shake like a baby's rattle. I am worried by how many there are and ask Farmer Martin if I should pull them out. He tells me not to bother: they will go, becoming fewer and fewer as the hedge fills out. One day, walking past them to the hut, I see that the thistles are covered with hundreds of meadow brown butterflies. They swarm into a cloud as I pass them and resettle on the thistles moments afterwards. There are so many I give up counting. Later I watch as dozens of goldfinches move across the meadow like a wave, settling on the thistle heads, which bob under their weight as they pluck seeds out with delicate curved beaks. They *pip, pip, pip* and light up the field with red-headed flashes of yellow, so that I understand their collective noun: a 'charm'. I wonder why, when they are so obviously important to so many creatures, I had been so keen to get rid of the thistles after all.

One day I see a grey heron fly over the fields, wings huge, legs dangling. It lands on a fence post alongside the scrub, next to the river by the white willow trees. It is not a rare bird but I have never seen it here before and my heart lifts sharply at how unusual I find its size and shape. Overhead the swifts and swallows pitch and soar. Once I see a sparrowhawk try to catch one. I watch the aerial battle, willing the swallow to escape. It does. Every day I worry I will go to the field and they will have left to start their long flight back to Africa, chasing their eternal summer. One day, I am right.

I cannot work out if everything around me is changing or if it is just that I am noticing it. The only thing I am sure has changed is me. I find myself thinking of a programme I once saw about an ancient oak tree at Kew Gardens in west London.[7] In 1987 a storm blew in from the south-west of England that felled fifteen million trees. In Kew Gardens, the 200-year-old Turner's Oak was heaved up by the wind and then dropped back into the earth at an angle. Before the storm the tree had been unwell and struggling to thrive for some time, although no one could work out why. They were sure that the ripping up of its roots in the storm would finally kill it off. They propped the old oak up with planks of wood so it was not a danger, then began to chainsaw the Gardens' other 700 fallen trees into pieces. There was another reason they left the old oak until last. Tony Kirkham, head of the Arboretum, was fond of it. The idea of having to cut it up was so sad that he wanted to put it off as long as he could. It took three years to clear all the other trees. But when they returned to the Turner's Oak they found it growing, strong and healthy, no longer sick. They realized that when the storm lifted the oak from its roots it pulled the soil apart. Air rushed in. When the oak fell back to earth it pushed air down into the soil making it porous, allowing oxygen and water to reach the roots more easily. The soil around the tree had become compressed by the weight of thousands of visitors. This was why the tree was dying. After the storm the tree put on more than a third of its growth. What happened to the Turner Oak changed the way that trees are looked after not just in Kew, but around the world. 'Trees are like people,' says Kirkham in the programme. 'They're moody. They stress. But they're beautiful when they're happy.' Now I think how our old life felt like it too was ripped up from its roots. As it turns out, it just needed some air.

We do mark harvest this year, with friends in the meadow field. Three months earlier it was filled with grazing ewes and their lambs. Now it is knee-high again. I pick flowers for the table – a few late buttercups, some daisies, lots of purple spiked knapweed alive with a carpet of bees feeding from them. The night is warm and everyone leaves in the early hours of the morning.

Afterwards, Ben and I lie on straw bales we put out for people to sit on. He immediately falls asleep. I lie down on the bale next to him and listen to the night around me. There is a barn owl hissing from the box in the oak tree to my right, sounding like static on the radio. Somewhere over the valley, a tawny owl hoots. As I lie under a tent of stars I see a flash of silver in the black sky above me. Then another. And another. I think about waking Ben – *Look, look, look: shooting stars everywhere!* – but something stops me, worried that no one's reaction can possibly reach what I am feeling right now. Later I learn I have been star bathing under the Perseid meteor shower caused by Earth slamming into debris left behind by a comet. After a long time I nudge Ben awake and we walk back across the field, slowly, lazily. As we do so, I realize something. Maybe it doesn't matter where I live. The connections I have formed are not necessarily specific to a certain place. They are to the world around me, wherever I may find it. Farmers were the ones who taught me this. They have taught me that everything is connected even when we do not notice the threads. They have taught me to look up. They have taught me how to see. And now that I have finally begun to, I realize I will never stop.

# Epilogue

At the beginning of the form I write: 'I know this may seem an unusual application for a forty-one-year-old mother of two small children to make . . .' It doesn't seem to matter. They let me in anyway.

On the first morning of my Graduate Diploma in Agriculture at the Royal Agricultural University in Cirencester it takes me forty-five minutes to get dressed. I'm worried that I do not own a checked shirt or a fleece gilet with leather piping or the right kind of trousers and will be thought unserious. In the end I just wear my usual clothes. No one seems to notice.

It starts because I want to learn how to drive a tractor. Mum says, 'Oh God, Sarah, I can teach you that,' but tractors now are very different from the ones she drove, so I look for a course instead. The tractor course somehow makes me click through to a one-year farming course, which somehow makes me click through to a diploma. Now I find myself in a room with sixteen strangers, learning about soil. My classmates range in age from twenty to over sixty. There is a paramedic, a documentary maker, two solicitors, a businessman suffering a self-diagnosed mid-life crisis, an army officer, a city estate agent, a mother-of-three accountant, a commercial surveyor, a Bulgarian agronomist and several others. There are as many women as there are men. Our teacher tells us we need to buy a lab coat. Next week we are going to dissect an animal organ. I am strangely excited.

My father came here over forty years ago to study farming. Now I am a student at the place he learned his trade. He finds this even funnier than I do. After he left, he became a land agent, and spent his days advising farmers and landowners on how best to manage their land, which mostly meant maximizing its productivity. He caused miles of hedges to be grubbed out, peat bogs drained, woods bulldozed and old, wild areas and native pasture to become edge-to-edge arable. There were huge government grants available to, as he will later describe it to me, 'wreck the countryside'. But as the years passed, he met some people who made him see things differently. Now he has 'crossed the Rubicon'. Crossing the Rubicon means he sends me emails like this.

From: Christopher Langford
To: Sarah Langford
Subject: Re: The Time is Now – Mass Lobby of Parliament
Sarah – I am going on a demo! This is on Wednesday 26 June at Parliament from 1:00 to 4:30PM
Are you going to be around that day if I was to call in?
DAD XXX

From: Sarah Langford
To: Christopher Langford
Subject: Re: The Time is Now – Mass Lobby of Parliament
Hilarious! For what??!
Yes please.
Xxx

From: Christopher Langford
To: Sarah Langford
Subject: Re: The Time is Now – Mass Lobby of Parliament
See you for coffee
The demo is really about your planet.

I had never really heard my father talk about the planet
before. But there he was – grey hair, walking shoes, sen-
sible sun hat – catching a train to London to join a
climate protest. Afterwards, we talk about how he feels,
in part, responsible. He did what he was paid to do and
what his clients wanted but, even so, he feels guilt. Now
the unintended consequences of industrial farming are
clear, I do not think he is the only one.

So while I know this diploma will teach me something, the last few years have shown me that the most
important lessons will come from people, not textbooks.
All over the country – all over the world – there are
farmers who have neither time nor inclination to write
down what they have learned from a life on the land.
They have learned these lessons from trying, failing,
watching, waiting and listening to those who were there
before them, weighing this advice against the feeling
they have in their heart and gut. Often this way of living
in the world has made philosophers of life from them –
even if they may not think of it this way – and so they
have much to contribute to our non-farming future also.

That is me. Here are the others.

On a cold, bleak day in winter the National Trust made
Charlie an offer he could not refuse. He can stay in the

farmhouse and keep a handful of nearby fields if he gives up arable farming and hands the rest of the land back. Before he knows it, the deal is done. Now he has security in the face of change and a chance of a different life. I understand why he accepts, even though, when I see him afterwards, I'm not sure who is more shocked by the decision: him or me. I email the National Trust and ask them what their plans for the farm are. Their reply talks about a change of land use. Woodland and hedgerow creation. Conservation grazing with native breeds. Food, it seems, will not feature. Later, Charlie tells me the quicker farmers like him pull out of food production, the faster the irresponsibility of rewilding and mass tree planting will show itself. He just hopes he lives long enough to see it. I look in the library for a copy of *Rural Rides* written by William Cobbett, the politician and farmer who rode around rural England between 1822 and 1826 to discover for himself the challenges farming communities faced. I wanted to see if, as Charlie had always said, the farm was in it. It was. 'These counties,' Cobbett wrote after riding through Cheriton and Kilmeston, where part of the farm still lies, 'are purely agricultural . . . Their hilliness, bleakness, roughness of roads, render them unpleasant to the luxurious, effeminate, tax-eating crew who never come near them . . .' No longer, then. For the first time in hundreds of years, 90 per cent of this farm will be used for leisure and not for growing food. Maybe this is the future of farming after all.

As a result of lockdown, in 2020 Ollie experienced a 50 per cent increase in the online sales of his meat from his farm shop. Sales the following spring and summer were up 20 per cent from the year before. He hopes that when his tenancy is up, and the council decides to sell the farm as their policy dictates they must, that he might be able to buy the farmhouse and the farm buildings. He hopes too that the council might make an exception to their policy to sell their farmland and offer him a new tenancy on it instead. He also has had a baby with his girlfriend. He is very, very tired. But happy.

Yolanda the cow died in her sleep on a straw bed on Tom's farm. She was twelve years old. Tom is still earning some of the best milk prices on the commercial market but the costs of fertilizers, fuel, feed and seed have gone up so much that, for the past six months, each litre of milk has cost him 3p more to make than he can sell it for. There are hidden costs too, like the price of haulage, and somehow these never seem to be passed on to the supermarket but always to the supplier. He is not entirely sure, with his dad now gone, that he can afford to see out the next decade. He sees lots of farmers around him selling up and getting out. The auctioneer at market told him he's never seen so many cows for sale before. Some of the farms that have been rented out for years are giving tenants' notice because they will make more money planting trees than grazing animals, even though some scientists warn that planting trees into peat soil does more harm than good. It doesn't matter. The landowners are following the money. On the fields around him, plastic tubes with tree whips in them are taking the place of cows.

The milk from Rebecca and Stuart's beloved Jersey

cows is now making cream, butter, yoghurt, ice-cream, fudge and ghee. The farm, shop and restaurant now have twenty-four full-time employees and more who work part-time and has started supplying Michelin-starred restaurants both in Norfolk and in London. They have planted a vineyard, which means they can now sell their own wine in their own farm shop. As well as the dairy herd and beef cattle, pigs, sheep and goats, they soon hope to start grazing free-range hens for meat on the pasture. In some of their fields they are paid extra to rest their land for a full year with nothing but stubble, then allow the beef cattle to graze off any greenery that has grown within it before new crops are drilled directly into the land in the spring. Last spring, for the first time, they calved the beef herd outside. Rebecca found that these native breeds needed no help at all with either birth or suckling. They just did it themselves. Although they are not officially organic, they have not used any artificial inputs on the farm for three years. They just haven't needed to.

George combined the organic regenerative wheat. He got a hell of a lot of clover seed but, mixed up within it, was enough wheat to sell. The hip urban bakery offered him £650 per tonne, which was over three times the price paid for milling wheat that year. Sam has read about a new/old perennial wheatgrass being developed by the Land Institute that might enable grain to be harvested without annual reseeding. He has already contacted those who are developing it for commercial use. It is not yet available for growers in the UK, they tell him, but when it is they assure him that Sam and

George – on a farm that others told them 'this won't work' – will be the first to know.

I am unsure what the long-term consequences of removing the subsidies that have supported food production for so long will be. Many are certain that this means farms will fail. I should be clear that not everyone thinks this is a bad thing. While few will say it with the record button pressed, many know someone whose poor farming has been covered up by public money. Some tell me these farms should fail so they can be worked by new blood with open minds, for there are plenty who want to do so. But if the food these new farmers make is undercut by imported products, made cheap because they have been produced without the same environmental and welfare standards, then this new generation of farmers will fail too.

With the price of land always increasing, the ability of these new farmers to buy or rent old farms is limited.[1] It seems therefore as likely that they may be bought not by a new farmer, but by those who do not intend to farm at all. This is already happening. In 2020 the amount of farmland bought by non-farmers accounted for just under half of all farm sales.[2]

In 1921, just over a hundred years ago, the Agriculture Act brought an end to the public money which had kept food production stable during the First World War. It removed guaranteed minimum wages and produce prices. Farming incomes dropped by up to 40 per cent within the year. At the same time, the government removed restrictions on cheap imported Canadian grain.

British farming was decimated. Rural poverty increased. Agricultural land was abandoned. When the Second World War began, just eighteen years afterwards, three quarters of all food consumed in Britain was being imported. The need for self-sufficiency became suddenly and urgently clear.

In 2020, a new Agricultural Act ended public money for food production, the legacy of which lay in that last war. At the same time, the government signed trade deals for tariff-free imports from large countries with lower food production standards on the other side of the globe. Now, as inflation begins to rise and the economy looks increasingly unstable, Russia invades Ukraine – two of the largest gas and grain producers in the world. It is hard not to see the parallels of a hat-trick of events a century apart. It is harder still not to wonder when we will begin to understand that food is not a public given, and we must become self-sufficient in ways which respect both food and the earth which grows it.

We are in a biodiversity and a climate crisis. Of that there is no doubt. Farming has played its part in that. The consequences of this are now apparent. I have never heard the song of a turtle dove because their numbers have fallen by 97 per cent in fifty years. I heard the call of a curlew and a lapwing for the first time when writing this book, because curlew numbers have nearly halved in twenty-five years and lapwings dropped by 80 per cent since 1960.[3] My children have never heard the sound of a cuckoo – a song my grandmother would pull me outside to listen to for she thought it the sound of spring beginning – because 65 per cent of all cuckoos have been lost in my lifetime. I try not to think too much about what birds my children's children will no longer hear.

But there are other crises too: of mental and physical heath, of housing and inequality which can forecast a child's future before it has even begun. And while our food may be cheap, the cost of dealing with this cheapness is not, whether that's the number of children admitted to hospital to have decaying teeth removed[4] or the cost in medication from weight-related illness.[5] The price of food should make hunger impossible, but it has not, because people's budgets have been decimated by reduced support and rising bills, including housing costs for homes built on land sold at an inflated price, the cost of which is then passed on to the home buyer, thanks to a rollover-tax loophole.[6] Nearly three million children in the UK live in households unable to afford enough food to meet official nutrition guidelines. Making food cheaper has not made people better off, financially or otherwise.

The solutions, it seems, may not be new ones. They can be found in books published a hundred years ago. If we connect food and health, both physical and mental, maybe governments might begin to see the benefits of properly funding the weekly shop for those families for whom food is still not cheap enough. If we connect food and the environment, our labelling and prices might start to show the true climate and environmental costs of its production. If we connect food to community, maybe we will not just give food its true place and value but see how all of us are connected to those who make it, and begin to value them too.

I have come to believe that farming can offer solutions to many of the problems we face. But, unexpectedly, I have also understood something else. Learning about farming taught me lessons about life applicable far beyond a field boundary. I learned about humility and patience and the unintended consequences that hubris can bring about. I learned about the need to find and trust your instincts. I experienced the power of mystery and awe in a time when both are often thought to be irrelevant. I saw how resilience meant diversity, especially so when different plants and animals grew alongside one another, and how, in life, diversity within our societies, businesses and public bodies enables strength and resilience in the same mutually beneficial way that animals, trees, plants, soil and microbes build a web of connectivity below and above the earth.

Finally, I came to understand that even if the 83 per cent of our population who live in urban areas do not know it or feel it, all of us are connected to our farmers, and them to us.[7] They may represent just 1 per cent of our workforce but they look after 70 per cent of our land. Their choices affect us all: whether we have clean air, clean water, rich wildlife, healthy soils in which to grow nutrient-rich food, habitats for the food chain we are part of, or whether our soils have the ability to store a pool of carbon three times larger than that which is in the atmosphere, an amount so vast it has the potential to cool our planet. Many have held on to a connection we have lost with the natural world. As we navigate a lonely online future some have learned lessons about life that even those who will never step foot in a field may find valuable. Farmers can do this. We just need to enable them. They carry both our country's future and its history in their hands. And, in doing so, they carry ours too.

# Acknowledgements

When I first dreamed up the idea for this book in 2019 farming was on few people's agendas. Even so, my power-house agent, the self-professed urbanite Nelle Andrews, let me persuade her both that farming really wasn't boring and that it was about to become A Thing. I am so grateful to her for always believing in me and for continuing to be enthusiastic, even when I talk about ploughing.

My editor, Tom Killingbeck, had the foresight to suggest I should not rush my research and writing, even if we both thought the virus that caused our first meeting to be on Zoom would all blow over in a few weeks. I am grateful for his patience, flexibility and insight, which have enabled this book to shapeshift into the one it has become, and to all the Penguin team who worked so hard upon it, particularly Jessamy Hawke for creating the most beautiful cover art out of nothing but a few photographs and ideas.

Without my parents-in-law's own desire to become farmers I might not have found my own. I owe them both a huge debt of gratitude for allowing us to take on the running of their farm and for their continued love and support.

My father, Christopher Langford, whose job I now understand for the first time, and my mother, Helen — and her siblings, Lizzie and Ian, allowed me to share their memories and diary entries of their life on the farm I loved so much. The greatest of thanks must go to my uncle, Charlie Flindt. We may not always agree on all the new-fangled nonsense, but he trusted me to tell the story of Hinton Ampner Farm, even if neither of us could foresee that it would end as it has.

I have neglected too many friends over the last few years, for which I am sorry, but especially my girlfriends, Kate Fortescue, Helia Ebrahimi, Amber Sainsbury and Olivia Rigg, whom I abandoned for a country life then returned to full of talk about soil. Thank you for your many decades of precious friendship, love and tolerance. Thank you also to Clover Stroud for her life-giving creative energy, support and ability to sit in silence alongside me as I write.

However, the most important people within this book are the farmers who shared their stories with me, especially Oliver and Philip White, Rebecca and Stuart Mayhew, Nick Padwick, Stephen Briggs, Martin Lines, AA and PB. It is hard to express just how much I have been humbled by your openness. All of you gave days of your life to me and let me into your hearts and minds so that I might better explain what you do and why you do it. You extended friendship as well as generosity. I hope this book has honoured that.

There were many people I visited and spoke to whose stories could not be crushed into the word count but my conversations and experiences with them left their imprint both on this book and on me. In particular I am very grateful to Chris Pratt, Chris Martin, Helen and

James Rebanks, James and Michelle Robinson, Tamara Hall, Sarah Bell, Henry Edmonds, Sam Burge, Paul Cherry, William Kendall, Tom Gribble, Ben Briggs, Jane Parker, George Dunn, Henry Dimbleby, Neil and Leigh Heseltine, Andrew and Rachel Knowles, Nick Van Cutsem, Nathan Nelson, Jason Borthwick, Megan Gimber, Claire Whittle, Julian Glover, Stephen Briggs and all my RAU classmates. Special thanks must also go to John Pawsey and Martin Durie, not only for being so generous with your time but for always so successfully shattering the stereotype of The Grumpy Farmer.

Finally, to Ben, who began this journey for us and without whom this book would not exist. My partner in every sense, thank you for showing me that farm-bureaucracy-dates can be quite special.

# Notes

## Introduction

1 The House of Commons Committee of Public Accounts, Environmental Land Management Scheme, 31st Report of Session 2021–22, p. 4.

## Chapter 1: The Beginning – Butter and Honey and Dust

1 Sam Knight, Swifts and the Fantasy of Escape, *New Yorker*, 26 July 2020.

## Chapter 2: Peter and Charlie

1 Approximately one third of English and Welsh Agricultural land is farmed by tenants, Barclay, Christopher. 'Tenant Farmers' Standard Note: SN/SC/1337. Available at: <https://researchbriefings.files.parliament.uk/documents/SN01337/SN01337.pdf> [Accessed 30 March 2022].
2 'By 1938 less than 4 per cent of the people made their living directly from the land, agriculture represented only 3.2 per cent of the national income, and industrial Britain imported two-thirds of its foodstuffs from abroad' (Alan F. Wilt, *Food for War: Agriculture and Rearmament in Britain before the Second World War*, Oxford University Press, 2001).

3 Figures from Save Our Magnificent Meadows. www.mag
nificentmeadows.org.uk/conserve-restore/importance-
of-meadows.

4 Adebayo Adeyinka, Erind Muco and Louisdon Pierre,
Organophosphates, National Center for Biotechnology
Information, 29 July 2021, ww.ncbi.nlm.nih.gov/books/
NBK499860/.

5 Plantlife, https://meadows.plantlife.org.uk/.

6 In the 1940s rural households provided themselves with
more than 92 per cent of their fruit and vegetables in
winter and 98 per cent in summer. Urban households
grew 12 per cent of their fruit and vegetables in winter
and 49 per cent in summer. About a third of the household income was spent on food in 1940 compared to
7 per cent now. DEFRA, The UK's Food History
Revealed Through Five Generations of Data, 1 September 2016.

7 From J. G. S. Donaldson and Frances Donaldson, *Farming
in Britain Today* (Penguin, 1969).

8 R. W. Wenlock, D. H. Buss, B. J. Derry and E. J. Dixon,
Household Food Wastage in Britain, *British Journal of
Nutrition*, vol. 43, no. 1, 1980, pp. 53–70.

9 Isabel Davies, Farmers Weekly 80th Birthday: Through
the Decades – 1980s, *Farmers Weekly*, 21 October 2014.

10 Ibid.

11 Ons.gov.uk. 2022, Family Spending in the UK – Office for
National Statistics. [online] Available at: <https://www.
ons.gov.uk/peoplepopulationandcommunity/personal
andhouseholdfinances/expenditure/bulletins/
familyspendingintheuk/april2019tomarch2020>.

12 www.ers.usda.gov/topics/international-markets-u-s-
trade/international-consumer-and-food-industry-
trends/.

13 Andrew O'Hagan, The End of British Farming, *London Review of Books*, vol. 23, no. 6, 22 March 2001; www.lrb.
co.uk/the-paper/v23/no6/andrew-o-hagan/the-
end-of-british-farming.

14 Some studies have claimed that regenerative farming systems provide both greater ecosystems and farmer profitability, generating 29 per cent lower production but 78 per cent higher profits than conventional systems. Claire E. LaCanne and Jonathan G. Lundgren, Regenerative Agriculture: Merging Farming and Natural Resource Conservation Profitably, *PeerJ*, 26 February 2018; https://peerj.com/
articles/4428/.

15 As a consequence of consuming cellulose, cattle burp (not fart) out methane, returning the carbon sequestered by plants back into the atmosphere, where it remains for ten years before being broken down and converted back into carbon dioxide, which is absorbed by plants, and so it goes round.

16 The carbon cycle dictates that whenever something dies and rots – tree, hedge, human, plant or animal – the carbon will be released again.

17 WWF and Tesco report 'Driven to Waste' quantified both food waste and food lost on farms globally.

18 S. Porter, D. Reay, E. Bomberg, and P. Higgins, 2018. Avoidable Food Losses and Associated Production-phase Greenhouse Gas Emissions Arising from Application of Cosmetic Standards to Fresh Fruit and

Vegetables in Europe and the UK, *Journal of Cleaner Production*, 201, pp. 869–78.

19 E. Holt-Giménez, A. Shattack, M. Altieri, H. Herren and S. Gliessman, 2012, We Already Grow Enough Food for 10 Billion People . . . and Still Can't End Hunger, *Journal of Sustainable Agriculture*, 36(6), pp. 595–8.

20 www.nationaltrust.org.uk/features/our-tree-planting-ambition;www.nationaltrust.org.uk/press-release/national-trust-outlines-fresh-ambition-in-landmark-speech-by-director-general.

21 The National Trust currently looks after a quarter of a million hectares of land, or 617,763 acres, and is the fourth biggest landowner in the UK and the second biggest in England. In 2016 the National Trust was criticized by the *Sunday Times* for selling off parcels of land for housing. The National Trust said it was obliged to sell the land to the highest-bidding developer because it was obliged to do so as a charity and one acre in every thousand it owns is earmarked for housing. As of 2020, 44 per cent of all the farmland owned by the National Trust is being managed in-hand. Lauren Harris, Why Farm Tenants are Criticising National Trust Landlords, *Farmers Weekly*, 9 October 2020; James Tapper, Priced-Out Tenants Accuse National Trust of Creating 'Ghost Villages', *Guardian*, 17 September 2016.

## *Chapter 3: Year One – Slow Walking*

1 Multiple studies undertaken over periods of up to thirty years that have followed the presence, decline and return

of raptors have concluded that raptors and corvids do not cause a decline in songbird numbers. See The Predation of Wild Birds in the UK: A Review of Its Conservation Impact and Management, Simon Barnes, 2008, Birds of Prey in the UK: On a Wing and a Prayer. Produced by the Royal Society for the Protection of Birds.

2 Anthropologist Professor Henry Bunn of Wisconsin University discovered evidence from a site in Tanzania showing that early humans were likely to have ambushed herds of animals up to 1.6 million years earlier than previously thought, pushing back the definitive date for the beginning of systematic human hunting by hundreds of thousands of years. H. Bunn and A. Gurtov, 2014, Prey Mortality Profiles Indicate that Early Pleistocene Homo at Olduvai Was an Ambush Predator, *Quaternary International*, 322–3, pp. 44–53.

3 The three definitions of sheep meat are:

Lamb – a young sheep under twelve months of age which does not have any permanent incisor teeth in wear.

Hogget – A term for a sheep of either sex having no more than two permanent incisors in wear, or its meat. Still common in farming usage, it is now rare as a domestic or retail term for the meat. Much of the 'lamb' sold in the UK is 'hogget' to a farmer in Australia or New Zealand.

Mutton – the meat of a female (ewe) or castrated male (wether) sheep having more than two permanent incisors in wear.

## Chapter 4: Ollie

1 John Bourn, BSE: The Cost of a Crisis, report by the Comptroller and Auditor General, National Audit Office, 8 July 1998 www.ncbi.nlm.nih.gov/pmc/articles/PMC7120640/.

2 Farmland has to be five hectares, or twelve acres or more to qualify for the basic payment scheme (BPS) subsidy.

3 The average farm size in the United Kingdom is 213 acres (86.2 hectares) (Sarah Dodds, What Size is the Average Farm?, *Macintyre Hudson*, 31 October 2019; www.macintyrehudson.co.uk/insights/article/what-size-is-the-average-farm).

4 11,300 farms of the country's largest farms account for 54 per cent of the country's farmland (ibid.).

5 www.pastureforlife.org/media/2018/10/PFL-Health-Benefits-at-14-Sept-FINAL.pdf.

6 Within their diet, chickens need to eat grain as well as greens. Free-range hens cannot survive off grass, but living outdoors does mean they can exhibit natural behaviours like scratching and pecking for worms and bugs.

   The American Pastured Poultry Producers' Association confirmed in studies that pastured poultry has significantly higher nutrient-rich meat, and eggs with greater levels of healthy omega 3 fatty acids and vitamins A and E than barn hens fed only on grain.

7 Joel Salatin, *You Can Farm: The Entrepreneur's Guide to Start and Succeed in a Farming Enterprise* (Polyface, 2006).

8 SI-UK; www.studyin-uk.com/study-guide/top-five-agriculture-universities-uk/.

9  Forage means food provided to grazing animals that they usually forage for themselves in fields or hedgerows, such as grasses, clovers, wildflowers, herbs and so on. As opposed to fodder, which means food conserved from the summer, such as hay or straw, usually fed to animals during the winter months.

10  By feeding on the muck, dung beetles dry it out faster, making it difficult for parasites like lungworm to lay their eggs in it, so breaking the worm cycle. On their shell backs, dung beetles carry tiny mites which use them to hitch a lift from cow pat to cow pat, where these mites feed on fly eggs, reducing the number which will later hatch and bother livestock. Now that many animals' muck is left in barns, not fields, and grassland is artificially fertilized, beetle numbers have dropped dramatically. One of the reasons for this is preventatively worming animals, with more wormer used as resistance is built up. But this thinking, says Dung Beetles For Farmers, is counter-productive: 'Beetles reduce parasite numbers. Wormers reduce beetle populations.' https://www.dungbeetlesforfarmers.co.uk/information-for-farmers.

11  In May 2020, the UK received less than 50 per cent of the average rainfall for that month. For England and Wales, it was the driest May since records began (in 1910) with 17 per cent of the average rainfall, and in some regions below 10 per cent. River flows across the UK had been dropping since mid-March, with some reaching their lowest flows for the time of year since records began.

12 Under the Protection of Badgers Act 1992, in England and Wales (the law is different in Scotland) it is an offence to:
- wilfully kill, injure or take a badger (or attempt to do so)
- cruelly ill-treat a badger
- dig for a badger
- intentionally or recklessly damage or destroy a badger sett, or obstruct access to it
- cause a dog to enter a badger sett
- disturb a badger when it is occupying a sett.

13 https://whiteoakpastures.wordpress.com/tag/live stock-guardian-dogs/.

14 Worldwide Opportunities on Organic Farms.

15 The Regenerative Organic Alliance – https://regenor-ganic.org/.

16 The Farm Retail Association found that, during the lockdown, 92 per cent of farm retailers reported a significant rise in new customers, 79 per cent of farm retailers introduced a click-and-collect service, 67 per cent introduced home deliveries and over 1.4 million orders were processed by farm shops for home delivery or collection. L. Harris, 2022, The Land Opportunities for Farmers in a Post-lockdown world, *Farmers Weekly*. [online] Available at: <https://www.fwi.co.uk/business/markets-and-trends/land-markets/where-are-the-opportunities-for-farmers-in-a-world-post-lockdown>.

## Chapter 5: Year Two – Looking Underneath

1 In medieval times, farming was based on open fields. These were cultivated by individual yeomen and tenant farmers

in disparate strips of land often placed far apart from one another. From as early as the twelfth century, however, agricultural land began to become consolidated into individually owned or rented fields through informal agreement as this was seen as a more economical way of farming.

However, during the seventeenth century Acts of Parliament, driven by landowners hoping to maximize rental income from their estates, consolidated these enclosures into law and, from the 1750s, enclosure by Act became common. Between 1604 and 1914 over 5,200 enclosure Bills were enacted by Parliament, enclosing just over a fifth of the total area of England, approximately 6.8 million acres.

These once large, communal open fields became hedged and fenced off and old boundaries disappeared. Historians remain divided over the extent to which enclosure forced those at the lowest end of rural society, the agricultural labourers, to leave the land permanently to seek work in the towns, with those historians in favour (mostly the beneficiaries) arguing that it was enclosure that enabled economic development, and those against arguing that it deprived the rural poor of both their livelihoods and common-law rights and was a driver for both urbanization and rural depopulation. 2022. [online] Available at: <https://www.parliament. uk/about/living-heritage/transformingsociety/ towncountry/landscape/overview/enclosingland/>.
2  The RSPB, 2022, A History of Hedges. [online] Available at: <https://www.rspb.org.uk/our-work/conservation/ conservation-and-sustainability/advice/conservation- land-management-advice/farm-hedges/history-

of-hedgerows/>. The 2007 Countryside Survey found that only 48 per cent of remaining hedgerows were in 'favourable condition'. In arable areas this was just 12 per cent.

3 J. Holden, R. Grayson, D. Berdeni, S. Bird, P. Chapman, J. Edmondson, L. Firbank, T. Helgason, M. Hodson, S. Hunt, D. Jones, M. Lappage, E. Marshal-Harries, M. Nelson, M. Prendergast-Miller, H. Shaw, R. Wade and J. Leake, 2019, The Role of Hedgerows in Soil Functioning within Agriculture Landscapes. Agriculture, Ecosytems&amp, *Environment*, 273, pp. 1–12.

4 Organic farming is based on enhancing the natural biological cycles in soil (e.g. nutrient cycling in the soil), in crops (e.g. encouraging natural predators of crop pests) and in livestock (e.g. the development of natural immunity in young animals); on building up soil fertility through the use of nitrogen fixation by legumes and enhancing soil organic matter; and on avoiding pollution. The aim is to work with natural processes rather than seek to dominate them, and to minimize the use of non-renewable natural resources such as the fossil fuels used for the manufacture of fertilizers and pesticides. Organic farming principles also encompass high animal-welfare standards and the improvement of the environmental infrastructure of the farm.

5 The British Poultry Council confirms that 3.5 per cent of the poultry market is free range and 1 per cent is organic. Indoor-reared chickens reach slaughter weight in 35–40 days, free-range chickens at 56 days and organic chickens at 81 days. R. Griffiths, and R. Griffiths, 2022. What is

338

Free-Range and Organic? – The British Poultry Council. [online]. Available at: <https://britishpoultry.org.uk/what-is-free-range-and-organic/>.

6 DEFRA monthly statistics on the activity of UK hatcheries and UK poultry slaughterhouses (data for December 2021). GOV.UK, Monthly Statistics on the Activity of UK Hatcheries and UK Poultry Slaughterhouses (data for February 2022). [online] Available at: <https://www.gov.uk/government/statistics/poultry-and-poultry-meat-statistics/monthly-statistics-on-the-activity-of-uk-hatcheries-and-uk-poultry-slaughterhouses-data-for-january-2021>.

7 Of the UK's 1,674 intensive farms, 85 per cent are poultry. Nearly 800 of them are big enough to fit the USA definition of a 'mega-farm' or 'Concentrated Animal Feeding Operation': at least 125,000 broilers, or 82,000 layers or pullets (chickens used for breeding), or 2,500 pigs, 700 dairy cattle or 1,000 beef cattle. Seven of the UK's ten largest poultry farms – producing meat or eggs or both – have the capacity to house more than a million birds. The biggest pig farm holds 23,000 pigs. The largest cattle farm holds 3,000 cattle. These are still modest compared to farms in Australia, America and elsewhere. The Bureau of Investigative Journalism (en-GB), 2022, The Rise of the "Megafarm": How British Meat is Made. [online] Available at: <https://www.thebureauinvestigates.com/stories/2017-07-17/megafarms-uk-intensive-farming-meat>.

# Chapter 6: Tom

1  www.theguardian.com/commentisfree/2010/feb/07/
   barabara-ellen-swine-flu.

2  https://healthcare-in-europe.com/en/news/european-
   parliament-to-investigate-who-pandemic-scandal.html.

3  www.nationalgeographic.com/science/article/
   experts-warned-pandemic-decades-ago-why-not-ready-
   for-coronavirus.

4  Department of Agricultural and Rural Affairs, February
   2016, United Kingdom Price Volume and Composition
   of Milk [online].

5  D. Bossio, S. Cook-Patton, P. Ellis, J. Fargione, J. Sander-
   man, P. Smith, S. Wood, R. Zomer, M. von Unger,
   I. Emmer and B. Griscom, 2020, The Role of Soil Carbon in Natural Climate Solutions, *Nature Sustainability*,
   3(5), pp. 391–8.

6  The number of cows decreased from 2.6 million in 1996 to
   1.9 million in 2020. The yield per cow, however, has
   increased by 100 per cent since 1975, from 4,100 to 8,200
   litres in 2020. Uberoi, E., 2021, UK Dairy Industry Statistics, [online] Available at: <https://researchbriefings.files.
   parliament.uk/documents/SN02721/SN02721.pdf>.

7  A 2015 YouGov poll found that 86 per cent of the public
   surveyed agreed 'that UK dairy cows should be able to
   graze on pasture and should not be permanently housed
   indoors'. However there is currently no compulsory
   labelling for milk. Unless shoppers choose organic milk
   they have no guarantee that their milk is from dairy herds
   able to graze outside.

8  In 2009, approximately 2,000 Holstein male dairy calves were killed each week (Compassion in World Farming Statistics). A 2018 report from the Cattle Health and Welfare Group – a UK cross-industry body made up of farm unions, levy boards and other experts – estimated that about one in five dairy bull calves were killed on farm. In 2020 dairy calf registrations were at the highest level for more than a decade, according to British Cattle Movement Service data, as some milk buyers contracted for male dairy calves to be reared rather than killed after birth. The practice of killing bull calves was banned under Red Tractor standards (the scheme applies to 95 per cent of milk produced in the UK) from the end of 2020. A cross-industry coalition is aiming to eliminate the killing of dairy calves on farm totally by 2023.

There have been numerous attempts to promote a market for high-welfare British veal (beef under a year old), championed by chefs such as Jamie Oliver and Jimmy Doherty but with little success, thanks to the legacy left by white veal, which was created by keeping calves in veal crates where they were fed only milk and no roughage to produce a lean, soft meat: a system banned in the UK in the 1990s. The RSPCA has campaigned to be allowed to rename veal as 'rosé beef' to end consumer misconception of it and create a market for dairy bull cows.

9  According to Compassion in World Farming, the use of sexed semen to prevent unwanted dairy bull calves has increased from 12 per cent in 2012 to over 30 per cent in 2019 and over 50 per cent in 2020. Over this same time

period, the percentage of male dairy calves born has more than halved. Compassion in World Farming, 2020, Progress for Male Dairy Calves. [online]

10 www.gov.uk/government/statistics/agri-climate-report-2021/agri-climate-report-2021.

11 D. Ryan Watts, 2019, Pollution: No River in England is Safe for Swimming. [online] Thetimes.co.uk.

## Chapter 7: Year Three – A Local Acre

1 Adjusting the tyre pressure as tractors move from road to field is proven to help lessen the impact of soil compaction: the softer the tyres, the less compaction. However, no matter how soft the tyres, increasing the weight of the tractor on top of them will still cause damage.

2 At time of writing, Zoopla listed the average sold price for a UK property as £329,291. In 2022, a John Deere four-track 8RX 370 costs £360,826 new, while a John Deere X9 1100 combine with 12.1m HDX header costs £850,328.

3 An average-sized tractor from 1980 was 75hp with an average weight of 2,850kg. An average-sized tractor from 2013 was 150hp with an average weight of 6,000kg, more than double the weight of the machines thirty years earlier. Theoretically, doubling the weight of a machine and doubling the tyre area that is in contact with the soil would maintain the same pressure, but in practice the overall weight of a machine in field conditions is rarely spread evenly. Figures taken from www.fwi.co.uk/machinery/is-farm-machinery-getting-too-heavy.

4 *Farmers Weekly*, 2014, Is Farm Machinery Getting Too Heavy?. [online] Available at: <https://www.fwi.co.uk/machinery/is-farm-machinery-getting-too-heavy>. For soil compression see: T. Keller, M. Sandin, T. Colombi, R. Horn, and D. Or, 2019, Historical Increase in Agricultural Machinery Weights Enhanced Soil Stress Levels and Adversely Affected Soil Functioning *Soil and Tillage Research*, 194, p. 104293. Around 3.9 million hectares are at risk of soil compaction in England and Wales – nearly twice the total area of Wales – with a potential yield penalty of £163 million every year, see: GOV.UK, 2022, United Kingdom Food Security Report, 2021. [online] Available at: <https://www.gov.uk/government/statistics/united-kingdom-food-security-report-2021>.

5 R. Douglas Brown, *East Anglia, 1939* (Terence Dalton, 1980).

6 Ibid., p. 92.

7 https://suffolkhorsesociety.org.uk/.

8 'They will beat their swords into ploughshares and their spears into pruning hooks' (Isaiah 2: 4). A bronze statue of a man hammering a sword into the shape of a ploughshare sits in the United Nations Headquarters in New York.

9 Plant blindness is defined as the inability to see or notice the plants in your environment. James Wandersee and Elisabeth Schussler, Preventing Plant Blindness, *American Biology Teacher*, vol. 61, no. 2, February 1999.

10 Increases in field sizes and a switch to crop monocultures have reduced the quantity and quality of habitats suitable for the hare, which favours a mosaic of varied crops, grassland and vegetation, both for food and shelter and to escape from predators. This impact has been

exacerbated by development: 1,121km² of arable farmland (an area the size of Bedfordshire) have been lost to urban development in Great Britain since 1990.

11 There were once 400 chemical companies in the UK. Now there is an oligopoly of five, with almost total control over the supply chains. In other parts of the world this domination is unusual: in Canada, for example, there are 265 distributor companies, which has limited their individual influence on the market. M. Dewes, 2019, What Can We Learn from Overseas to Better Curate the Use of Pesticides. [online] Nuffieldscholar.org.

12 Ibid.

13 J. Smith, B. Pearce and M. Wolfe, 2012, A European Perspective for Developing Modern Multifunctional Agroforestry Systems for Sustainable Intensification, *Renewable Agriculture and Food Systems*, 27(4), pp. 323–32.

14 Www1.montpellier.inra.fr, 2022, Welcome to the SAFE Homepage : Silvoarable Agroforestry for Europe. [online] Available at: <https://www1.montpellier.inra.fr/safe/english/index.htm>. Details of an ongoing twelve-year UK project can be found here: Innovativefarmers.org, 2021, Twelve year field lab into the benefits of silvopasture launched. [online] Available at: <https://www.innovativefarmers.org/news/2021/february/18/twelve-year-field-lab-into-the-benefits-of-silvopasture-launched/>.

15 Jay Griffiths, *Kith: The Riddle of the Childscape* (Hamish Hamilton, 2013).

16 *How the Edwardians Spoke*, BBC Four, July 2007.

1 The industry standard for mortality is hard to pinpoint as pre-weaning deaths aren't included. Most units will aim for post-weaning mortality of around 5 per cent, but Rebecca estimates that, if the whole cycle is taken into account (i.e. birth to death), the figure is nearer 12 to 15 per cent.

2 Liebig's experiments, and those of other agricultural scientists afterwards, took place in laboratories, greenhouses and small field plots, although they rarely examined the soil at depth.

3 Living organisms present in soil include archaea, bacteria, fungi, algae, protozoa, and a wide variety of larger soil fauna including springtails, mites, nematodes, earthworms, ants and other insects that spend all or part of their life underground. Soil fungi are also large components of the soil. A. Fortuna, 2012, The Soil Biota, *Nature Education Knowledge*, 3(10):1.

4 D. H. McNear Jr., (2013), The Rhizosphere – Roots, Soil and Everything in Between, *Nature Education Knowledge*, 4(3):1.

5 www.fwi.co.uk/livestock/grassland-management/a-guide-to-mob-grazing-livestock.

6 The rhizosphere is the thin area of soil immediately surrounding the root system. Plant-root exudates are fluids emitted through the roots of plants. These secretions influence the rhizosphere around the roots, inhibiting harmful microbes and promote the grow of self and kin plants. Y. Zhu, X. Lin and H. Chu, 2022. Editorial:

Rhizosphere Microbiome Special Issue, *Plant and Soil*, 470(1–2), pp. 1–3.

7 A hallmark of 'Norfolk husbandry' in the early eighteenth century was typically a six-year rotation of crops to prevent exhaustion of the soil and remove the need for a fallow year because a crop of turnips, sown in rows, could be hoed to remove the weeds that were usually cleared by ploughing after leaving the land fallow. The rotation was soon adopted by other farms. Coke of Norfolk then strengthened the idea of rotation with his famous saying, 'No two white straw crops one after the other.' In the popular Norfolk four-course rotation wheat was grown in the first year, turnips in the second, followed by barley – with clover and ryegrass undersown – in the third. The clover and ryegrass were grazed or cut for feed in the fourth year. The turnips were used for feeding cattle and sheep in the winter. The system worked because the fodder crops eaten by the livestock produced large supplies of previously scarce animal manure, which in turn was richer because the animals were better fed. When the sheep grazed the fields, their waste fertilized the soil, promoting heavier cereal yields in following years. The system became common on the newly enclosed farms by 1800, remaining almost standard practice on most British farms for the best part of the following century. During the first three-quarters of the nineteenth century, it was adopted in much of continental Europe.

8 In 1750 the population of England was approximately 5.7 million. It had probably reached this level before, during the Roman period, then again in 1300 and 1650 but had

not grown beyond it as agriculture could not respond to the pressure of feeding more people. However, the population then grew from 5.7 million in 1750 to 16.6 million in 1850 because of changes in agriculture. This was partly due to new rotational farming systems but also land reclamation, including the draining of the peat fenlands, the clearing of woodland and the enclosure of upland pastures from the seventeenth century onwards. From the mid-seventeenth century farmers began to grow clover to fix nitrogen, as well as beans, peas and vetches. The new system of farming was remarkable because it was sustainable while also vastly increasing the output of food per agricultural worker, and this improved agricultural productivity in turn enabled the industrial revolution to flourish. By 1850 only 22 per cent of the British workforce was employed in agriculture, the smallest proportion in the world. Mark Overton, *Agricultural Revolution in England: The Transformation of the Agrarian Economy, 1500–1850* (Cambridge University Press, 1996).

## Chapter 9: Year Four – Examining Roots

1 The University of Edinburgh, 2018, A Third of Fruit and Veg Crop Too Ugly to Sell. [online] Available at: <https://www.ed.ac.uk/research/latest-research-news/a-third-of-fruit-and-veg-crop-too-ugly-to-sell>.

2 Estimates suggest soil degradation, erosion and compaction result in losses of about £1.2 billion each year and reduce the capacity of UK soils to produce food. Intensive agriculture has already caused arable soils to lose 40 to

60 per cent of their organic carbon; 2 million hectares are at risk. GOV.UK, 2022, United Kingdom Food Security Report 2021. [online] Available at: <https://www.gov.uk/government/statistics/united-kingdom-food-security-report-2021>.

3 In 2020 the organic UK market grew by 12.6 per cent. Net farm income increases for organic cropping farms, resulting in a significantly higher figure when compared to non-organic (£690/ha compared to £288/ha). While organic cropping farms incurred a 14 per cent greater total cost per hectare, variable costs were £86/ha less, with fertilizer and crop protection contributing most to this difference. Soilassociation.org, 2021, Organic Farming and Growing – Does it Stack Up? Market Update, 2021. [online] Available at: <https://www.soilassociation.org/farmers-growers/market-information/organic-farming-and-growing-does-it-stack-up/>.

The Rodale Institute in Pennsylvania, which has undertaken the longest running side-by-side comparison of conventional and organic methods over some forty years, found that after a fifty-year transition period organic yields were competitive with conventional yields in good years and significantly outperformed them in drought or flood conditions. The organic systems used 45 per cent less energy, released 40 per cent fewer carbon emissions and earned three to six times higher profits for farmers; Rodaleinstitute.org, 2011, The Farming Systems Trial – Celebrating 30 Years. [online] Available at: https://rodaleinstitute.org/wp-content/uploads/fst-30-year-report.pdf.

4 Jones, L., *Losing Eden: Why Our Minds Need the Wild* (Penguin Publishing Group, 2021).

5 A crop which has been desiccated is easier to harvest and the seed contains less moisture so requires less drying. When the widely used glyphosate-based herbicide RoundUp was developed by Monsanto, its application pre harvest was never considered.

6 D. Goulson, 2013, Neonicotinoids and Bees: What's All the Buzz?. *Significance*, 10(3), pp. 6–11. A. Harrison-Dunn, 2021, Why Are Banned 'Bee-Killer' Neonicotinoids Still Being Used in Europe?. [online] Modern Farmer. Available at: <https://modernfarmer.com/2021/03/why-are-banned-bee-killer-neonicotinoids-still-being-used-in-europe/>.

7 B. Phillips, A. Navaratnam, J. Hooper, J. Bullock, J. Osborne and K. Gaston, 2021, Road Verge Extent and Habitat Composition Across Great Britain, *Landscape and Urban Planning*, 214, p. 104159.

8 P. Greenfield, 2020, On the Verge: A Quiet Roadside Revolution is Boosting Wildflowers. [online] *Guardian*. Available at: <https://www.theguardian.com/environment/2020/mar/14/on-the-verge-a-quiet-roadside-revolution-is-boosting-wildflowers-aoe>.

9 2013. YouGov Woodland Trust Survey Results. [online] Available at: https://d25d2506sfb94s.cloudfront.net/cumulus_uploads/document/048vpv8neq/YG-Archive-Woodland-Trust-survey-results-240613-native-trees.pdf.

10 R. Lines-Kelly, 2000, Soil Sense. Wollongbar: NSW Agriculture.

11 A study over ten years concluded that earthworm totals were significantly higher with disc-harrowed land compared to ploughed land. Indeed, all tillage treatments that did not involve ploughing resulted in higher species richness. The research concluded that loss of earthworm biodiversity had a negative effect on soil functioning in agricultural ecosystems. G. Ernst and C. Emmerling, 2009, Impact of Five Different Tillage Systems on Soil Organic Carbon Content and the Density, Biomass, and Community Composition of Earthworms after a Ten-year Period, *European Journal of Soil Biology*, 45(3), pp. 247–51.

12 L. Jones, *Losing Eden: Why Our Minds Need the Wild* (Penguin Publishing Group, 2021).

13 It costs approximately £1 per sheep to have wool sheared. A fleece can be sold for 70p per kilo. The average UK fleece tends to weight between two and four kilos. Farmers have been reduced to burning or composting wool instead of paying to pack and transport it.

14 Online Etymology Dictionary. [online] Available at: <https://www.etymonline.com/>.

15 The Environment Agency's natural capital register and accounts tool and scorecard can be requested from the EA's Natural Capital Team. See their website: <https://eftec.co.uk/project/environment-agency-natural-capital-register-and-account-tool>.

16 I. Napper, B. Davies, H. Clifford, S. Elvin, H. Koldewey, P. Mayewski, K. Miner, M. Potocki, A. Elmore, A. Gajurel, and R. Thompson, 2020, Reaching New Heights in Plastic Pollution – Preliminary Findings of Microplastics on

Mount Everest, *One Earth*, 3(5), pp. 621–30. X. Peng, M. Chen, S. Chen, S. Dasgupta, H. Xu, K. Ta, M. Du, J. Li, Z. Guo and S. Bai, 2018, Microplastics Contaminate the Deepest Part of the World's Ocean, *Geochemical Perspectives Letters*, pp. 1–5.

17 As well as offering vital habitats to a variety of creatures, research has shown that trees are often effective at filtering pollutants from the air, dispersing carbon dioxide, nitrogen dioxide and particulate matter from the atmosphere, and becoming both the liver and lungs of an ecosystem. K. Beckett, P. Freer-Smith and G. Taylor, 2000, The Capture of Particulate Pollution by Trees at Five Contrasting Urban Sites. *Arboricultural Journal*, 24(2–3), pp. 209–30.

18 M. Aizen, L. Garibaldi, S. Cunningham and A. Klein, 2009. How Much Does Agriculture Depend on Pollinators? Lessons from Long-term Trends in Crop Production. *Annals of Botany*, 103(9), pp. 1579–88.

19 T. Šantl-Temkiv, M. Sahyoun, K. Finster, S. Hartmann, S. Augustin-Bauditz, F. Stratmann, H. Wex, T. Clauss, N. Nielsen, J. Sørensen, U. Korsholm, L. Wick and U. Karlson, 2015, Characterization of Airborne Ice-nucleation-active Bacteria and Bacterial Fragments, *Atmospheric Environment*, 109, pp. 105–17.

20 Instituteforgovernment.org.uk, 2022, Agriculture Subsidies after Brexit. [online] Available at: https://www.instituteforgovernment.org.uk/printpdf/10673.

21 The RSPB, n.d., Skylark Threats – The RSPB. [online] Available at: <https://www.rspb.org.uk/birds-and-wildlife/wildlife-guides/bird-a-z/skylark/threats/.

1 In 1981 the world's record wheat yield was in Scotland and produced 13.99 tonnes per hectare (5.66 per acre). England beat the world wheat record in 2015 after harvesting 16.52 tonnes per hectare (6.69 per acre). K. Fletcher, 2020, Scottish Crop Master Chases Holy Grail of Wheat Yields. [online] The Scottish Farmer. Available at:https://www.thescottishfarmer.co.uk/news/18663652. world-beating-wheat-yields-former-record-holder-gordon-rennie-tells-us-achieve/. Since the millennium, however, wheat yields in the UK have remained mostly stagnant. Our World in Data, n.d., Crop Yields Data Explorer. [online] Available at: https://ourworldindata. org/explorers/crop-yields?tab=chart&facet=none&cou ntry=~GBR&Crop=Wheat&Metric=Actual+yield.

2 Pigs are no longer allowed to be fed on food scraps.

3 *Farmers Weekly*, 2000, Deloitte & Touche Report – analysis. [online] Available at: https://www.fwi.co.uk/news/ deloitte-touche-report-analysis.

4 The State of Nature Report. [online] Available at: <https://nbn.org.uk/wp-content/uploads/2019/09/ State-of-Nature-2019-UK-full-report.pdf>.

5 Rothamsted Research has been tracking and studying all types of bugs in the UK since 1964 making it the longest-standing insect monitoring facility of its kind in the world. It has over eighty traps spread around the country to sample insect populations every day.

 A review of seventy-three of the most comprehensive studies into insect decline concluded that 40 per cent of insect species are declining, with the total mass of insects falling by 2.5 per cent every year. Climate change is likely to affect insect populations, but the most notable impact was from increased pesticide use and habitat loss. G. Gooderham, 2018, The Sussex Study: 50 Years of Monitoring an Agricultural Ecosystem. [online] Gwct.org.uk. Available at: <https://www.gwct.org.uk/media/1244244/National Geographic_SussexStudy50.pdf.

In the State of Nature Report 2019, a similar pattern of decline is evident for butterflies in the wider countryside. There is growing concern about pollinators, largely related to use of pesticides such as neonicotinoids but also the decrease in plant diversity and flower-rich habitats.

6 According to the Intergovernmental Panel on Climate Change (IPCC), synthetic nitrogen fertilizers have increased by 800 per cent since the 1960s. Research shows that in most intensive agriculture systems, over 50 per cent and up to 75 per cent of nitrogen applied is not used by plants and is lost by leaching into the soil. B. Hirel, T. Tétu, P. Lea and F. Dubois, 2011, Improving Nitrogen Use Efficiency in Crops for Sustainable Agriculture, *Sustainability*, 3(9), pp. 1452–85. This nitrogen washes into, lakes, oceans and rivers. It causes algae to grow unnaturally fast, using up all available oxygen in the water and suffocating other species creating 'dead zones' of algae blooms.

Research shows that synthetic fertilizers are responsible for 1 out every 40 tonnes of GHGs pumped into the atmosphere. S. Menegat, A. Ledo and R. Tirado, 2021, Greenhouse Gas Emissions from Global Production and Use of Nitrogen Synthetic Fertilisers in Agriculture.

7 *Farmers Weekly*, 2000, Stores Cite UK Organic Food Shortage. [online] Available at: <https://www.fwi.co.uk/news/stores-cite-uk-organic-food-shortage>.

8 Cirencester is the home of the Royal Agricultural University.

9 Eve Balfour, *The Living Soil* (Faber & Faber, 1943).

10 Albert Howard, *An Agricultural Testament* (Oxford University Press, 1940).

11 Lord Northbourne, lifelong farmer and lecturer in agriculture at the University of Oxford, first used the term 'organic' in his book *Look to the Land* (Dent, 1940) to describe the kind of farming that had 'a biological completeness' (p. 86).

12 A. O'Hagan, *The End of British Farming* (Profile Books, 2001).

13 Heritage wheats are older varieties of wheat that were grown before the chemical revolution. They aren't as old as 'ancient grains' like millet and sorghum and have been selectively bred over generations. They tend to grow taller than modern wheats, making them prone to 'lodging' (falling over), and do not produce as great a yield as modern wheat varieties.

## Chapter 11: Year Five – Paying Attention

1 P. Newton, N. Civita, L. Frankel-Goldwater, K. Bartel and C. Johns, 2020, What Is Regenerative Agriculture? A Review of Scholar and Practitioner Definitions Based on Processes and Outcomes, *Frontiers in Sustainable Food Systems*, 4.

2 According to a recent study, the use of cover crops across 85 per cent of annually planted US cropland could sequester around 100 million tons of carbon dioxide per year. This would offset about 18 per cent of US agricultural-production emissions and 1.5 per cent of total US emissions. Fargione et al., 2018, Natural Climate Solutions for the United States, Science Advances. [online] Available at: <https://www.science.org/doi/pdf/10.1126/sciadv.aat1869>.

3 While 84 per cent of glyphosate applied is absorbed by the soil, 16 per cent remains soluble and, during rainfall, is washed deeper into the soil's profile, where it is unable to biodegrade (T. Shushkova, I. Ermakova and A. Leontievsky, Glyphosate Bioavailability in Soil, *Biodegradation*, vol. 21, no. 3, 2009, pp. 403–10). Surface runoff then transports the chemical into water systems. In another 2009 study it was detected in 36 per cent of a large sample of water sources from states in the US Midwest. Glyphosate application led to increased soil concentrations of nitrate by 1,592 per cent and phosphate by 127 per cent, increasing the risk of nutrient leak into water systems (Mailin Gaupp-Berghausen et al., Glyphosate-Based Herbicides Reduce the Activity and Reproduction of Earthworms and Lead to Increased Soil Nutrient Concentrations, *Scientific Reports*, vol. 5, 5 August 2015; www.nature.com/articles/srep12886).

4 Research has confirmed that, in laboratory conditions, glyphosate transfers from the leaves and into the rhizosphere around a plant's roots, where it enters the soil system to be re-absorbed by non-target crops. This has

also been shown to cause a reduction in root growth (R. Kanissery et al., Glyphosate: Its Environmental Persistence and Impact on Crop Health and Nutrition, *Plants*, vol. 8, no. 11, 2019, p. 499).

Glyphosate's interference with plant growth inhibits a crop's ability to absorb micronutrients such as magnesium, zinc, iron and boron, all of which are critical to establishing disease resistance (G. Neumann et al., Relevance of Glyphosate Transfer to Non-Target Plants Via the Rhizosphere, *Journal of Plant Diseases and Protection*, vol. 20, 2006, pp. 963–9).

It also blocks the plant from synthesizing key amino acids by disrupting an enzyme critical for the shikimate pathway that can leave crops vulnerable to attack by soil-borne pathogens, further increasing their vulnerability to disease (G. Johal and J. Rahe, Glyphosate, Hypersensitivity and Phytoalexin Accumulation in the Incompatible Bean Anthracnose Host-Parasite Interaction, *Physiological and Molecular Plant Pathology*, vol. 32, no. 2, 1988, pp. 267–81).

Glyphosate was found to be a leading factor in the development of diseases in wheat and barley developed from the *Fusarium* pathogen; disease was found to be highest in crops under no-till management (C. Fernández-Quintanilla et al., Is the Current State of the Art of Weed Monitoring Suitable for Site-Specific Weed Management in Arable Crops? *Weed Research*, vol. 58, no. 4, 2018, pp. 259–72).

Glyphosate has been linked to a crop's ability to access phosphorus as it competes for absorption by soil particles with this mineral (S. Eker et al., Foliar-Applied Glyphosate Substantially Reduced Uptake and Transport of Iron and Manganese in Sunflower (*Helianthus annuus* L.) Plants, *Journal of Agricultural and Food Chemistry*, vol. 54, no. 26, 2006, pp. 10019–25; A. Gimsing, A. and O. Borggaard, Competitive Adsorption and Desorption of Glyphosate and Phosphate on Clay Silicates and Oxides, *Clay Minerals*, vol. 37, no. 3, 2002, pp. 509–15).

It has been identified as a key factor in nutrient deficiencies in crops owing to its ability to interfere with the uptake and translocation of these nutrients throughout a crop (Kanissery et al., Glyphosate).

5 A tine weeder is a large agricultural rake mounted in rows and pulled behind a tractor, used to remove weeds from fields.

6 M. Abram, 2020, How Regenerative Farming Cut Fixed Costs by 40% in First Year – *Farmers Weekly*. [online] Available at: https://www.fwi.co.uk/arable/land-preparation/soils/how-regenerative-farming-cut-fixed-costs-by-40-in-first-year?share=telegram.

7 U. Sezen, 2020, The Ancient Oak Tree that Taught the World a Lesson – BBC (2020). [online] Nature Documentaries. Available at: <https://naturedocumentaries.org/19185/ancient-oak-tree-taught-world-lesson-bbc/>.


## Epilogue

1 Savills.co.uk, 2021, The Outlook for Farmland Values in Great Britain. [online] Available at: <https://www.savills.co.uk/research_articles/229130/309959-0>.

2 Strutt & Parker's Farmland Database states that farmland purchased by lifestyle buyers and private investors is at 47 per cent of all sales, its highest ever level.

3 All figures taken from the RSPB or the British Trust for Ornithology.

4 The primary reason for NHS hospital admissions amongst children aged five to nine years old is teeth extraction. For the financial year 2019 to 2020 the estimated cost of extraction operations because of tooth decay in children up to nineteen years old was £33 million, a reduction from previous years. GOV.UK, 2021, Hospital Tooth Extractions of 0 to 19-year-olds. [online] Available at: https://www.gov.uk/government/publications/hospital-tooth-extractions-of-0-to-19-year-olds. The dentistry industry describes these extractions as 'avoidable'. G. Bissett, 2020, Tooth Decay Remains Most Common Reason for Hospital Admission in Five to Nine-Year-olds. [online] https://dentistry.co.uk/.

5 A 2019 study amongst women aged fifty-five to seventy-nine years old found that excess weight accounted for an estimated 11 per cent (£229 million) of all consultation costs and 20 per cent (£384 million) of all medication costs. S. Kent, S. Jebb, A. Gray, J. Green, G. Reeves, V. Beral, B. Mihaylova and B. Cairns, 2018, Body Mass Index and Use and Costs of Primary Care Services Among Women Aged 55–79 Years in England: A Cohort and Linked Data Study, *International Journal of Obesity*, 43(9), pp. 1839–48.

In the government paper Tackling Obesity, published 27 July 2020 (www.gov.uk/government/publications/

tackling-obesity-government-strategy/tackling-obesity-empowering-adults-and-children-to-live-healthier-lives), it was estimated that two-thirds of adults are overweight and half of these are living with obesity; one in three children leave primary school overweight and one in five are already living with obesity. Obesity is highest amongst the most deprived groups in society and is directly linked with reduced life expectancy and disease.

The annual NHS spend on the treatment of obesity and weight-related diabetes is greater than the amount spent on the police, the fire service and the justice system combined. These costs are projected to reach £9.7 billion per year by 2050. GOV.UK, 2017, Health Matters: Obesity and the Food Environment. [online] Available at: <https://www.gov.uk/government/publications/health-matters-obesity-and-the-food-environment/health-matters-obesity-and-the-food-environment--2>.

6  Rollover relief is an exemption on paying capital gains tax if money earned from the sale of farmland is reinvested in 'qualifying assets' (buying more farmland, buildings and fixed plant or machinery) within three years. A landowner can therefore sell farmland for development and pay no CGT if they buy more farmland with the sale money within three years. As farms generally only come up for sale in spring or autumn, this provides an incentive to overpay to ensure the purchase is completed in time and tax avoided. Land prices rise accordingly. This is then passed on to the buyer of those houses built on farmland.

In 2017 the Office for National Statistics published for the first time estimates for the aggregate value of the land in the UK. It showed that in two decades, the market value of land quadrupled.

7 DEFRA, *Statistical Digest of Rural England*, https://assets.publishing.service.gov.uk/government/uploads/system/uploads/attachment_data/file/1028819/Rural_population__Oct_2021.pdf.